Inquiry Based Learning Gui

Chemistry

TENTH EDITION

Steven S. Zumdahl
University of Illinois at Urbana-Champaign

Susan Arena Zumdahl
University of Illinois at Urbana-Champaign

Donald J. DeCoste
University of Illinois at Urbana-Champaign

Prepared by

Donald J. DeCoste
University of Illinois at Urbana-Champaign

Australia • Brazil • Mexico • Singapore • United Kingdom • United States

ISBN: 978-1-305-95749-7

Cengage Learning
20 Channel Center Street
Boston, MA 02210
USA

Cengage Learning is a leading provider of customized learning solutions with office locations around the globe, including Singapore, the United Kingdom, Australia, Mexico, Brazil, and Japan. Locate your local office at: **www.cengage.com/global**.

Cengage Learning products are represented in Canada by Nelson Education, Ltd.

To learn more about Cengage Learning Solutions, visit **www.cengage.com**.

Purchase any of our products at your local college store or at our preferred online store **www.cengagebrain.com**.

Printed at CLDPC, USA, 11-16

TABLE OF CONTENTS

Preface

Inquiry learning is based on the premise that students will better understand the material if they are active participants in the learning process, not just passive sponges for information. Dictionary definitions of inquiry include aspects such as questioning, a search for knowledge, deliberation, and interrogation. Basically, to inquire is to ask and consider questions, and this requires the students to be engaged in the material.

The questions and activities in the *Inquiry Based Learning Guide (IBLG)* are designed to get students to consider the underlying concepts involved in understanding chemistry. We want the students to recognize that learning chemistry goes beyond "getting the right answer" to algorithm-based exercises. We want them to learn to think like chemists—to be able to solve problems because they truly understand the underlying concepts, not because they have memorized a solution to a particular type of problem.

Your teaching experience has probably shown (and there has been a great deal of research to back this up) that students can be successful at solving even relatively difficult math-based chemistry problems yet still not have a firm understanding of the chemical principles involved in the problem. This is partly due to the fact that students generally solve these problems by using memorized rules, algorithms, or equations. While we as instructors mean for the problems to be conceptual in nature, students often exchange thinking about the chemistry with finding a shortcut to solve a specific problem. Because of this, many of the questions in the *IBLG* explicitly ask the students to explain a term, equation, rule, or algorithm. In other words, we ask the students to explain what we have long been assuming they had to know to solve the traditional problems.

How does the *Inquiry Based Learning Guide* help students learn chemistry?
There are two main impediments to learning that the *IBLG* addresses. The first consists of content misconceptions, and the second is the way in which many students define teaching and learning.

The questions and activities in the *IBLG* have been specifically written to surface students' initial ideas and content misconceptions. Students (and instructors) often do not realize the misconceptions harbored by the students until the students have to explain their ideas about specific topics. Once these misconceptions come to light, they can be confronted, and conceptual change can take place.

While these content misconceptions undoubtedly play a role in students' lack of understanding, there is an even more fundamental difficulty: misconceptions about what it means to really understand. For example, we have found:

1. **Students are passive readers of textbooks.** They often view the textbook as simply a place to find their homework assignments, and perhaps to look for sample problems and solutions. Even students who take the time to read some of the text will often skip over graphs and figures, which are crucial aids for increasing conceptual understanding. Many of the questions in the *IBLG* send the students back to the text to consider specific statements, or to discuss the significance of a table, graph, or figure.

2. **Even successful students rely more on math ability than problem-solving approaches.** Although at the University of Illinois we have some of the brightest math students in the country, many of these students are not very proficient at thinking through problems in chemistry. We have had many students make claims such as, "Once I am shown how to set up a problem I can usually solve it," and genuinely believe that this demonstrates understanding. It is not so much that they are trying to get out of thinking, but that they are defining understanding in terms of algorithms. If we give them a problem without an obvious algorithm, they struggle. The problems in the *IBLG* ask the students to consider the chemistry in the system, and to reflect on how they solve the problems.

How do the questions in the *Inquiry Based Learning Guide* address these concerns?

As stated previously, the essence of the questions in the *IBLG* is to ask the students explicitly to discuss the fundamental ideas of chemistry and to consider how they solve problems. These problems generally require the students to explain rather than calculate; even when the students are asked to calculate, they are also asked to explain their reasoning and the significance of the answer. In this way, the students are forced to confront misconceptions, and they learn to actively seek answers (the nature of inquiry) and evaluate their problem-solving abilities. Let's consider a few sample problems from the *IBLG* and discuss their rationale.

Example #1: Relating microscopic to macroscopic

Question: Provide microscopic drawings down to the atoms for Figure 1.16 in your text.

Rationale: This problem not only sends students to the text, it also has the students consider their understandings of concepts such as *element*, *compound*, and *mixture* and to relate them to the microscopic world. One of the reasons chemistry is so difficult for students is that we live in the macroscopic world, but discuss chemistry in the microscopic world. It is a good idea to get the students to relate these two as early and often as possible.

Example #2: Explicitly considering a term

Question: Why is the term *sodium chloride molecule* incorrect but the term *carbon dioxide molecul*e correct?

Rationale: This question has the students consider what the term *molecule* means, and they must decide when it does apply and when it does not. It also addresses a misconception, because since we can write the formula NaCl, many students believe it is a molecule.

Example #3: Explicitly considering a procedure

Question: On page 92 of the text, in describing how to determine a formula, it states the "easiest way to do this is to work with 100.00 grams of the compound."

a. Why is this the "easiest way"?
b. Could we assume any mass? Prove your answer using a homework problem.
c. Could we assume 1.0 mole of the compound? Support your answer.

Rationale: This problem not only sends students to the text, it also has the students explain why it is they solve a particular problem in a certain way, and consider if this is the only way to solve it.

Example #4: A math problem with no numbers

Question: Estimate the mass of air at normal conditions that takes up the volume of your head. Provide support for your answer.

Rationale: This problem requires the students to determine (and estimate) the data required to solve a problem, and then to solve it. This requires the students to explicitly think about the approach they will use to solve the problem. The goal is to help the students develop their own strategies for solving the problems.

Example #5: What if…?

Question: What if energy was not conserved? How would this affect our lives?

Rationale: This problem requires the students to discuss the significance of an idea (in this case energy conservation) by imagining what would happen if it were not true. In framing a question this way, the students are required to think more deeply about the topic.

How should the *Inquiry Based Learning Guide* be used?

A crucial part of learning involves true discussion of the concepts using questions that require the students to probe their understanding of the concepts. As instructors, we know that trying to teach something to someone is one of the surest ways to expose our misunderstandings of the underlying concepts. Peer teaching is one of the best features of group discussions. However, for group work to yield real value, the questions must be carefully chosen. For example, while relatively traditional quantitative problems have their place in a chemistry course, they usually do not elicit much in the way of discussion. The main goal of traditional problems is usually to get the correct answer (although the process is also important, the emphasis is usually on the answer). That is not to say that one type of problem is more important than the other, but that both types need to be provided to the students. The questions in the *IBLG* come from our own experiences with increasing the level of discussion and peer teaching at the University of Illinois.

The ideal way to use these questions is to have relatively small groups of students (3 or 4) working together during class time. This serves many purposes, including:

1. **The students are effectively modeling how to think about problems.** Many ideas surface, debates occur, students often go down "wrong paths," and they sometimes discuss important issues that are not explicitly written into the questions. This type of environment also teaches the students the benefits of actively thinking through a problem.

2. **Students are required to be active participants in their own education.** This is important for reasons already mentioned and one more—we have found that, many times, students need to say something out loud before they can really evaluate it. For example, a student who believes he/she understands a concept (but does not) will often start out

speaking quickly and confidently and then begin to trail off as he/she "listens" to what he/he is saying.

3. **You can interact with the students and question them further.** This is important to keep students from using shortcuts to just get a correct answer, and to question the students even when they are correct (to keep them from being passive acceptors of answers). The answers we accept from the student are as important as the questions we ask of them.

4. **You get to know the students better more quickly.** This includes a better understanding of how they approach problems, their misconceptions, and group dynamics.

Where can I find the solutions to the problems in the *IBLG*?

Solutions to the problems and activities in the *IBLG* will be posted on the online Instructor's Companion Site. Hints about students' conceptions of certain topics and how to use the problems are also included where appropriate.

A final note of caution and encouragement

We hope you will find the questions and activities in the *Inquiry Based Learning Guide* useful in your courses. That said, while having the students take a more active role in the learning process is a long-term goal, the students often struggle with this initially. Students, especially those who have been successful algorithm solvers in the past, will often become easily frustrated with having to explain their ideas and their approaches to problem solving. However, the more you can show the growth in learning that these students experience, the more they will accept that this approach is a better one. One of the greatest compliments we have heard from one of our students is, "I am trying to learn how to ask myself the same questions that you ask us. Then I will not need you anymore." This is the true essence of life-long learning and surely a goal of any teacher.

Chemistry: An Overview and The Scientific Method

1. In Section 1.1 of your text, the terms *macroscopic* and *microscopic* are used. Explain what is meant by each of these terms and give an example of each.

2. What is our current microscopic view of the world? What evidence do we have to support this view?

3. How does taking a microscopic view help to explain chemical reactions?

4. As stated in the text, there is no one scientific method. However, making observations, formulating hypotheses, and performing experiments are generally components of "doing science". Read the following passage, and list any observations, hypotheses, and experiments. Support your answer.

   Joyce and Frank are eating raisins and drinking ginger ale. Frank accidentally drops a raisin into his ginger ale. They both notice that the raisin falls to the bottom of the glass. Soon, the raisin rises to the surface of the ginger ale, and then sinks. Within a couple of minutes, it rises and sinks again. Joyce asks "I wonder why that happened?" Frank says, "I don't know, but let's see if it works in water." Joyce fills a glass with water and drops the raisin into the glass. After a few minutes, Frank says, "No, it doesn't go up and down in the water." Joyce closely observes the raisins in the two cups and states, "Look there are bubbles on the raisins in the ginger ale but not on the raisins in the water." Frank says "It must be the bubbles that make the raisin rise." Joyce asks, "OK, but then why do they sink again?"

5. Here is an activity for you to try. Add water to a beaker or glass so that it is about ¾ full. Without disturbing the water, carefully add a couple of drops of food coloring.
   a. Make careful observations of the water. What happens to the food coloring? Describe your observations in some detail.
   b. Develop a hypothesis for your observations that takes a microscopic perspective (see Section 1.1).
   c. Design experiments to determine what effect the temperature of the water has on the results. Carry out these experiments, make observations, and develop a hypothesis for your observations.

6. Explain the difference between a *theory* and a *law*. Provide an example of each.

7. Many people mistakenly believe that if a theory has enough support (or if it is believed for a long period of time), it becomes a law. Explain why this is not true.

8. Give an example of when you have used the scientific method outside of school. Explain this example.

Uncertainty, Measurement, and Calculations

1. Your text makes the following claim: It is very important to realize that a *measurement always has some degree of* ***uncertainty***.

   Why does a measurement always have to have some degree of uncertainty? Give an example measurement and discuss the uncertainty in the measurement.

2. Look at Figures 1.8a, 1.8b, and 1.8c in your text. Explain why each is labeled as it is (for example, Figure 1.10a is labeled "neither accurate nor precise").

3. The text does not include a figure labeled "accurate but not precise". Is this even possible? If yes, provide a figure and explain it. If no, explain why not.

4. Sketch two pieces of glassware showing the proper graduations: one that can measure with a precision of ±0.1 mL and one that can measure with a precision of ±0.01 mL

5. The beakers below have different precisions as shown.

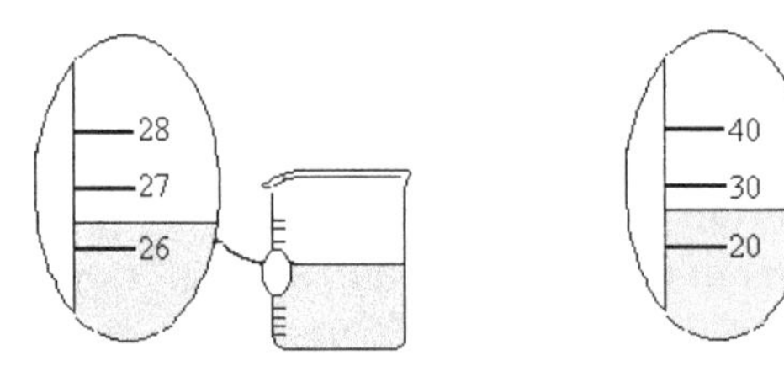

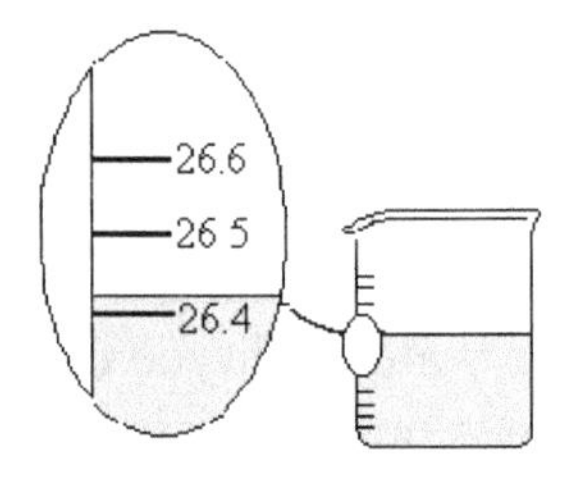

   a. Label the amount of water in each of the three beakers to the correct number of significant figures.
   b. Suppose you pour the water from these three beakers into one container. What should be the volume in the container reported to the correct number of significant figures?

6. True or False? For any mathematical operation performed on two measurements, the number of significant figures in the answer is the same as the least number of significant figures in either of the measurements.

   Explain your answer.

7. You can decide which digits are significant in a measurement by converting that measurement to scientific notation. Explain why this is true and provide a few examples.

8. Complete the following and explain each in your own words: leading zero's are (never/sometimes/always) significant, captive zero's are (never/sometimes/always) significant, and trailing zero's are (never/sometimes/always) significant.

   For any with an answer of "sometimes", give examples of when the zero is significant and when it is not and explain.

9. Do the *Measuring a Book* activity. You will find the instructions at the end of this chapter.

10. What is the numerical value of a unit factor? Why must this be true?

11. You are planning to tour the country in your car over the summer. To make sure you can afford it, you have to estimate the amount of money you'll spend on gasoline to make such a trip. Discuss what you will need to know to make this estimation, provide reasonable values for these numbers, and make a sample calculation.

12. Determine the number of seconds in a month. Do this in at least two ways using different unit factors. Compare your answers.

Temperature and Density

1. Look at Figure 1.9 in your text. At which temperature do the Fahrenheit and Celsius temperatures have the same numerical value? Prove that this is true using either Equation 1.1 or Equation 1.2 in your text.

2. Consider three samples of water at the same temperature all experiencing an increase in temperature. The temperature of sample A increased by 1°C, that of sample B increased by 1°F, and that of sample C increased by 1K. Rank the samples in order of increasing final temperature and support your answer.

3. In order to understand an equation, it is useful to be able to derive it. The equation relating the Fahrenheit and Celsius temperature scales is derived in Section 1.8. Review this derivation and then do problem 64 at the end of Chapter 1.

4. Density can be a difficult concept because it is a ratio of mass to volume. People sometimes claim that a heavier object must be more dense than a lighter object, although this is not necessarily the case. Problem 107 at the end of Chapter 1 is a good problem to do in order to understand this ratio. Do problem 107.

5. Look at Table 1.5 in your text. How do the densities of gases, liquids, and solids compare to each other? Use microscopic pictures to explain why this is true.

Classification of Matter

1. Explain the terms *element*, *atom*, and *compound.* Provide an example and microscopic drawing of each.

2. Mixtures can be classified as either homogeneous or heterogeneous. Compounds cannot be classified in this way. Why not? In your answer, explain what is meant by heterogeneous and homogeneous.

3. Provide microscopic drawings down to the atoms for Figure 1.14 in your text.

4. Use the following figures to answer the questions below.

a)

b)

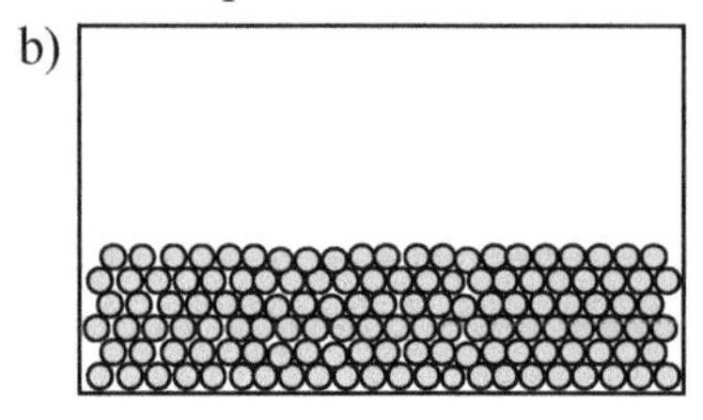

c)

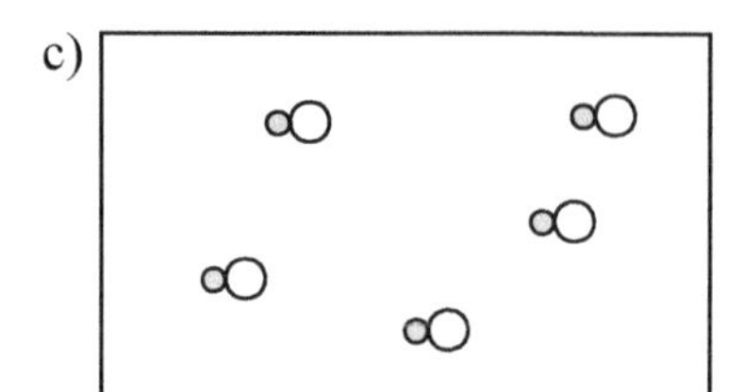

d)

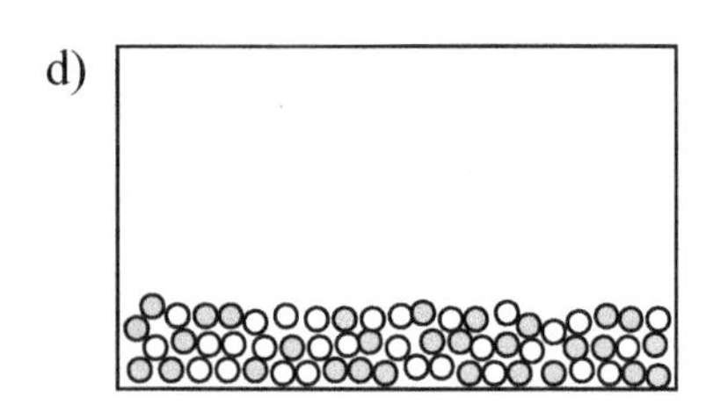

e)

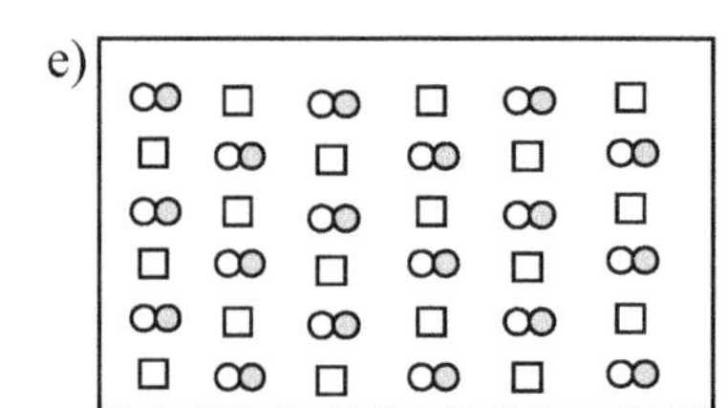

a. Which best represents a homogeneous mixture of an element and a compound?

b. Which best represents a gaseous compound?

c. Which best represents a solid element?

d. Which best represents a heterogeneous mixture of two elements?

e. What would you term the choice not chosen in a, b, c, or d?

Measuring a Book

Materials
- 4 special rulers
- Chemistry textbook

Procedure
1. Measure the length and width of your chemistry textbook with each of the four special rulers. Record these values.

2. Convert all measurements to centimeters (if necessary). Record these values.

3. Using your measurements, calculate the perimeter (in cm) and area (in cm^2) of the cover of your chemistry textbook. Record these values.

Analysis and Conclusions
1. Which ruler gives you the most precise measure of the perimeter and area of the cover of your chemistry textbook? Why?

2. Which ruler gives you the least precise measure of the perimeter and area of the cover of your chemistry textbook? Why?

3. Justify the number of significant figures in each of your measurements.

4. Justify the number of significant figures in each of your calculations.

5. Compare your measurements with other groups. For which ruler was there the most difference among groups? The least difference? Explain.

6. Compare your calculations with other groups. For which ruler was there the most difference among groups? The least difference? Explain.

Summary Table
Record your measurements and calculations in the table below.

Ruler #	length	width	length	width	perimeter	area
1	ft	ft	cm	cm	cm	cm^2
2	in	in	cm	cm	cm	cm^2
3	cm	cm	cm	cm	cm	cm^2
4	cm	cm	cm	cm	cm	cm^2

Development of the Atomic Theory

1. Explain how the law of definite proportions and the law of multiple proportions support Dalton's atomic theory.

2. Explain how Avogadro's hypothesis supports existence of diatomic hydrogen, oxygen, and chlorine.

3. Science often develops by using the known theories and expanding, refining, and perhaps changing these theories. In Section 2.4 it was seen that Rutherford used Thompson's ideas when thinking about his model of the atom. What if Rutherford did not know about Thompson's work? How might Rutherford's model of the atom been different?

4. Rutherford was surprised when some of the alpha-particles bounced back. He was surprised because he was thinking of Thompson's model of the atom. What if Rutherford believed atoms were as Dalton envisioned them? What do you suppose Rutherford would have expected and what would have surprised him?

The Nature of the Atom

1. It is good practice to actively read the text book and to try to verify claims that are made when you can. These problems ask you to do this.

   a. The following claim is made in your text: "…a piece of nuclear material about the size of a pea would have a mass of 250 million tons."

      Provide mathematical support for this statement.

   b. Consider Figure 2.14 in your text. The caption of the photo on page 54 in your text states that if an atomic nucleus the size of a ball bearing, a typical atom would be the size of a football stadium.

      Provide mathematical support for this statement.

2. Use Figure 2.14 in your text as a reference for this question. If the average size of an atom was the same size as a ball bearing, about how tall would you be? Provide mathematical support for your answer.

3. Using the data in Table 2.1 in your text, estimate the mass of electrons in your body. Provide mathematical support for your answer.

4. Suppose you could see atoms. How would an atom of carbon be similar to an atom of oxygen? How would they be different?

5. Differentiate between an atomic element and molecular element. Provide an example and microscopic drawing of each.

6. Both atomic elements and molecular elements exist. Are there such entities as atomic compounds and molecular compounds? If so, provide an example and microscopic drawing. If not, explain why not.

7. Now that you have gone through Chapter 2, go back to Section 2.3 and review Dalton's Atomic Theory. Which of the premises are no longer accepted? Explain your answer.

8. How is an ion formed?

9. A certain ion has 27 electrons and a charge of 2+. Write the symbol for the ion.

10. Why is the term "sodium chloride molecule" incorrect but the term "carbon dioxide molecule" is correct?

Nomenclature

1. The rules for naming compounds are very systematic. As your text mentioned, initially there was no system for naming compounds and it would be impossible to learn the common names of all of the chemical compounds. Do the *Naming Compounds* activity. This activity will help you understand the rules for naming compounds as well as appreciate how the rules are systematic and why the rules exist as they do. You will find the instructions at the end of this chapter.

2. In some cases the Roman numeral in a name is the same as a subscript in the formula, and in some cases it is not. Provide an example (formula and name) for each of these cases. Explain why the Roman numeral is not necessarily the same as the subscript.

3. The formulas $CaCl_2$ and $CoCl_2$ look very similar. What is the name for each compound? Why do we name them differently?

4. The formulas MgO and CO look very similar. What is the name for each compound? Why do we name them differently?

5. Explain how you use the periodic table to tell you that there are two chloride ions for every magnesium ion in magnesium chloride and one chloride ion for every sodium ion in sodium chloride. Then write the formulas for calcium oxide and potassium oxide and explain how you got them

6. What is the general formula for an ionic compound formed by elements in the following groups? Explain your reasoning and provide an example for each (name and formula).
   a. Group 1 with group 7
   b. Group 2 with group 7
   c. Group 1 with group 6
   d. Group 2 with group 6

7. An element forms an ionic compound with chlorine having the formula XCl_2. The ion of element X has mass number 89 and 36 electrons. Identify the element X and tell how many neutrons it has.

8. Explain any problems with each of the given names. Then, identify the formulas for the compounds with the given name (there may be more than one possible answer) and provide the systematic name for this compound (name each of the possible formulas from an incorrect name)

name	problem	formula	systematic name
barium dichloride			
carbon oxide			
copper (II) sulfate			
iron oxide			
diphosphorus pentoxide			
potassium sulfide			
perchloric acid			
sulfur hexafluoride			
magnesium phosphide			
calcium (II) nitrate			

9. It will be helpful when considering problems in solutions for you to think about what the solutions "look like" at a particulate level. Answer the following questions concerning ionic compounds dissolved in water.

   a. If you dissolve 10 units of iron (III) chloride, there will be _______ particle(s) in solution. The formula for iron (III) chloride is __________________.

   b. If you dissolve 5 units of $Pb(NO_3)_2$, there will be _______ particle(s) in solution. The name of $Pb(NO_3)_2$ is ______________________.

   c. If you dissolve 8 molecules of HCl, there will be _______ particle(s) in solution. The name of HCl is _____________________.

   d. If you dissolve 7 units of $K_2Cr_2O_7$, there will be _______ particle(s) in solution. The name of $K_2Cr_2O_7$ is ______________________________.

   e. If you dissolve 6 units of calcium hydroxide, there will be _______ particle(s) in solution. The formula for calcium hydroxide is _________________.

   f. If you dissolve 3 units of Cs_2SO_3 there will be _______ particle(s) in solution. The name of Cs_2SO_3 is ______________________________.

   g. If you dissolve 2 units of barium phosphate, there will be _______ particle(s) in solution. The formula for barium phosphate is ___________________.

   h. If you dissolve 1 unit of NH_4CN, there will be _______ particle(s) in solution. The name of NH_4CN is _____________________.

   i. If you dissolve 4 units of copper (I) chloride, there will be _______ particle(s) in solution. The formula for copper (I) chloride is _______________.

   j. If you dissolve 4 units of copper (II) chloride, there will be _______ particle(s) in solution. The formula for copper (II) chloride is _______________.

Naming Compounds

1. Given the following names and formulas, and using a periodic table, devise a set of rules for naming chemical compounds.

carbon dioxide	CO_2	carbon monoxide	CO
sulfur trioxide	SO_3	sodium chloride	$NaCl$
iron(III) chloride	$FeCl_3$	dinitrogen monoxide	N_2O
sulfur hexafluoride	SF_6	magnesium chloride	$MgCl_2$
potassium oxide	K_2O	iron(II) sulfide	FeS
oxygen dichloride	OCl_2	aluminum oxide	Al_2O_3

2. Using the rules you developed, name the following compounds.

   a. $CaBr_2$ ____________________

   b. CCl_4 ____________________

   c. BeO ____________________

   d. $CoCl_2$ ____________________

   e. NI_3 ____________________

3. Using the rules you developed, provide formulas given the following names.

   a. lithium sulfide ____________________

   b. copper(I) oxide ____________________

   c. phosphorus trichloride ____________________

   d. strontium chloride ____________________

   e. iron(III) oxide ____________________

Counting By Weighing and Atomic Masses

1. Do the *Counting Pennies without Counting* activity. You will find the instructions at the end of this chapter.

2. Why do we need to count atoms by weighing them?

3. True or false? Most hydrogen atoms have a mass of 1.008 amu.

   Justify your answer. That is, if true, explain why it is true. If false, what mass do most hydrogen atoms have?

4. Copper consists mainly of two isotopes, ^{63}Cu and ^{65}Cu. Which is more abundant? How do you know?

The Mole and Molar Mass

1. Explain the concept of the mole in your own words. How do we use it and why is it useful? Discuss how the mole concept allows us to use the average atomic masses on the periodic table and how it aids in counting by weighing.

2. This claim is made in your text: 1 mole of marbles is enough to cover the entire earth to a depth of 50 miles.

   Provide mathematically support for this claim. Is it reasonably accurate?

3. Estimate the length of time it would take you to count to Avogadro's number. Provide mathematical support.

4. Estimate the number of atoms in your body and provide mathematical support. Remember, it is an estimate, so although you should choose your numbers wisely, they need not be exact.

5. Consider separate equal mass samples of magnesium, zinc, and silver. Rank these from greatest to least number of atoms and support your answer.

6. You have a 20.0g sample of silver metal. You are given 10.0 g of another metal and told that this sample contains twice the number of atoms as the sample of silver metal. Identify this metal.

7. Do the *Relative Masses* Activity. You will find the instructions at the end of this chapter.

8. Explain the difference between the terms *atomic mass* and *molar mass.*

9. A molecule has a mass of 4.65×10^{-23} g. Provide two possible chemical formulas for such a molecule.

10. How would you find the number of "ink molecules" it takes to write your name on a piece of paper with your pen? Explain what you would need to do and provide a sample calculation.

11. Consider separate samples of water and carbon dioxide, each with the same mass. Which contains the greater number of molecules? How many times greater?

12. Consider separate equal mass samples of each of the following:

    H_2O N_2O $C_3H_6O_2$ CO_2

    Rank these from greatest to least number of oxygen atoms and support your answer.

Percent Composition of Compounds and Empirical and Molecular Formulas

1. Consider separate equal mass samples of each of the following:

    H_2O N_2O $C_3H_6O_2$ CO_2

    Rank these from highest to lowest percent oxygen by mass and support your answer.

2. True or false: The atom with the largest subscript in a formula is the atom with the largest percent by mass in the compound.

    If true, explain why with an example. If false, explain why and provide a counterexample. In either case, provide mathematical support.

3. Vinegar is a mixture of acetic acid ($HC_2H_3O_2$) and water. Glucose ($C_6H_{12}O_6$) is a product of photosynthesis. Find the percent by mass composition of acetic acid and glucose. How do your results indicate that percent composition data is incomplete for determining the formula of a compound? What else do you need?

4. A substance A_2B is 60.0% A by mass. Calculate the %B by mass for AB_2.

5. You find two compounds that contain only carbon and oxygen. The first is 27.0% carbon by mass, and the second is 43.0% carbon by mass.

    Do these data contradict the law of multiple proportions because 43 is not a multiple of 27? Prove your answer.

6. Provide at least 2 examples of molecules that have the same empirical and molecular formulas.

7. For which of the following is the percent by mass of carbon equal to the percent by mass of oxygen? Justify your choice.
    a) CO
    b) $C_8H_{15}O_3$
    c) $C_2H_4O_2$
    d) $C_4H_6O_3$

Balancing Chemical Equations

1. How is the balancing of chemical equations related to the law of conservation of mass?

2. Consider the chemical equation $H_2 + O_2 \rightarrow H_2O$. Why cannot we balance the equation as follows: $H_2 + O_2 \rightarrow H_2O_2$? Use molecular level pictures like those in Section 3.7 to support your answer.

3. At the beginning of this chapter you learned about counting by weighing. Now that you have learned about balanced chemical equation, why must we do this? That is, why must we know how many atoms or molecules we have?

4. Which of the following correctly describes the balanced chemical equation given below? There may be more than one true statement. If a statement is incorrect, explain what is incorrect about it.

$$4Al + 3O_2 \rightarrow 2Al_2O_3$$

   a. For every 4 atoms of aluminum that react with 6 atoms of oxygen, 2 molecules of aluminum oxide are produced.
   b. For every 4 moles of aluminum that react with 3 moles of oxygen, 2 moles of aluminum (III) oxide are produced.
   c. For every 4 grams of aluminum that react with 3 grams of oxygen, 2 grams of aluminum oxide are produced.

5. Which of the following correctly balances the chemical equation given below? There may be more than one correct balanced equation. If a balanced equation is incorrect, explain what is incorrect about it.

$$CaO + C \rightarrow CaC_2 + CO_2$$

   I. $CaO_2 + 3C \rightarrow CaC_2 + CO_2$
   II. $2CaO + 5C \rightarrow 2CaC_2 + CO_2$
   III. $CaO + (2.5)C \rightarrow CaC_2 + (0.5)CO_2$
   IV. $4CaO + 10C \rightarrow 4CaC_2 + 2CO_2$

6. Which of the following are true concerning balanced chemical equations? There may be more than one true statement.

   I. The number of molecules is conserved.
   II. The coefficients tell you how much of each substance you have.
   III. Atoms are neither created nor destroyed.
   IV. The coefficients indicate the mass ratios of the substances used.
   V. The sum of the coefficients on the reactant side equals the sum of the coefficients on the product side.

7. Consider the generic chemical equation $aA + bB \rightarrow cC + dD$ (where a, b, c, and d represent coefficients for the chemicals A, B, C, and D respectively).
   a. How many possible values are there for "c"? Explain your answer.
   b. How many possible values are there for "c/d"? Explain your answer.

Reaction Stoichiometry

1. It can be easy to get "lost in the math" with stoichiometric calculations. You should never lose sight of what is happening in the chemical reaction. To make sure you are thinking about the molecules and equating these with the symbols of the chemical equation, answer Active Learning Questions 6 and 7 at the end of Chapter 3.

2. An individual coefficient in a balanced chemical equation doesn't really tell us anything. Explain why this is true.

3. What is meant by the term *mole ratio*? Give an example of a mole ratio and explain how it is used in solving a stoichiometry problem.

4. Which would produce a greater number of moles of product: a given amount of hydrogen gas reacting with an excess of oxygen gas to produce water, or the same amount of hydrogen reaction with an excess of nitrogen gas to make ammonia? Support your answer.

5. Methane (CH_4) reacts with oxygen in the air to produce carbon dioxide and water.

   Ammonia (NH_3) reacts with oxygen in the air to produce nitrogen monoxide and water.

   What mass of ammonia is needed to react with excess oxygen in order to produce the same amount of water as 1.00 g of methane reacting with excess oxygen?

6. Consider a reaction represented by the following balanced equation

$$2A + 3B \rightarrow C + 4D$$

   You find that it requires equal masses of A and B so that there are no reactants left over. Which of the following is true? Justify your choice.
   a) The molar mass of A must be greater than molar mass of B.
   b) The molar mass of A must be less than molar mass of B.
   c) The molar mass of A must be the same as molar mass of B.

7. Baking powder is a mixture of cream of tartar ($KHC_4H_4O_6$) and baking soda ($NaHCO_3$). When it is placed in an oven at typical baking temperatures (as part of a cake, for example), it undergoes the following reaction (the CO_2 makes the cake rise):

$$KHC_4H_4O_6(s) + NaHCO_3(s) \rightarrow KNaC_4H_4O_6(s) + H_2O(g) + CO_2(g)$$

   You decide to make a cake one day, and the recipe calls for baking powder. Unfortunately, you have no baking powder. You do have cream of tartar and baking soda, so you use stoichiometry to figure out how much of each to mix.

   Of the following choices, which is the best way to make baking powder? The amounts given in the choices are in teaspoons (that is, you will use a teaspoon to measure the baking soda and cream of tartar). Justify your choice.

Assume a teaspoon of cream of tartar has the same mass as a teaspoon of baking soda.

a) Add equal amounts of baking soda and cream of tartar.
b) Add a bit more than twice as much cream of tartar as baking soda.
c) Add a bit more than twice as much baking soda as cream of tartar.
d) Add more cream of tartar than baking soda, but not quite twice as much.
e) Add more baking soda than cream of tartar, but not quite twice as much.

Calculations Involving a Limiting Reactant

1. Consider a reaction in which two reactants make one product (for example, consider the unbalanced equation $A + B \rightarrow C$). You know the following:

   2.0 moles of A (with an excess of B) can make a maximum of 2.0 moles C
   3.0 moles of B (with an excess of A) can make a maximum of 4.0 moles C

   If you react 2.0 moles of A and 3.0 moles of B, what is the maximum amount of C that can be produced?

2. Consider a chemical equation with two reactants forming one product. If you know the mass of each reactant, what else do you need to know to determine the mass of the product? Why isn't the mass necessarily the sum of the mass of the reactants? Provide a real example of such a reaction and support your answer mathematically.

3. Consider the balanced chemical equation

$$A + 5B \rightarrow 3C + 4D$$

   When equal masses of A and B are reacted, which is limiting? Justify your choice.

   a) If the molar mass of A is greater than the molar mass of B, then A must be limiting.
   b) If the molar mass of A is less than the molar mass of B, then A must be limiting.
   c) If the molar mass of A is greater than the molar mass of B, then B must be limiting.
   d) If the molar mass of A is less than the molar mass of B, then B must be limiting.

4. Which of the following reaction mixtures would produce the greatest amount of product, assuming all went to completion? Justify your choice.

   Each involves the reaction symbolized by the equation:

$$2H_2 + O_2 \rightarrow 2H_2O$$

   a) 2 moles of H_2 and 2 moles of O_2.
   b) 2 moles of H_2 and 3 moles of O_2.
   c) 2 moles of H_2 and 1 moles of O_2.
   d) 3 moles of H_2 and 1 moles of O_2.
   e) Each would produce the same amount of product.

5. Given the equation: $3A + B \rightarrow C + D$, if 4.0 moles of A is reacted with 2.0 moles of B, which of the following is true? Justify your choice.

   a) A is the limiting reactant because 3 moles of A react with every 1 mole of B.
   b) A is the limiting reactant because you need 6 moles of A and have 4.
   c) B is the limiting reactant because you have fewer moles of B than A.
   d) B is the limiting reactant because 3 moles of A react with every 1 mole of B.
   e) Neither reactant is limiting.

6. Do the *Nuts and Bolts of Stoichiometry* activity. You will find the instructions at the end of this chapter.

Counting Pennies without Counting

Materials

- sealed container with pennies
- empty container similar to the sealed container
- ten pennies
- balance

Procedure

1. Obtain a sample of pennies in a sealed container. Do not count the pennies.

2. Obtain an empty container that is similar to the one containing the pennies and ten pennies.

3. Devise a method using a balance to determine the number of pennies in the sealed container without opening it. Only use one of the pennies. Determine this number.

4. Repeat step 3 using all ten pennies instead of one penny. Use your method to determine the number of pennies in the sealed container.

5. Open the sealed container and count the pennies.

Results/Analysis

Which method (using one penny or ten pennies) allowed you to more accurately determine the number of pennies in the sealed container? Why? How does this finding relate to counting atoms by weighing?

Relative Masses

Materials

- cotton balls
- paper clips
- rubber stoppers
- balance

Procedure

1. Obtain cotton balls, paper clips, and rubber stoppers from your teacher.
2. Devise a method to find the average mass of each of these objects. Discuss this method with you teacher.
3. Determine the average mass of each object.

Results/Analysis

1. Copy the table below onto your paper.
2. Record the average masses of each of the objects in your table.
3. Give the lightest object a relative mass of 1.0 in your table. Determine the relative masses of the remaining objects and record these in your table.
4. How many of the lightest object would you need to have a pound of that object? Call this number n.
5. If you had n of each of the other objects, how much would each sample weigh? Fill these numbers into your table.
6. Determine which columns correspond to the chemical terms "atomic mass" and "molar mass."
7. Which number represents "Avogadro's number?"

	Mass of one object	Relative mass of one object	Mass of n objects
Cotton balls	g		lb.
Paper clips	g		lb.
Rubber stoppers	g		lb.

The Nuts and Bolts of Stoichiometry

Materials

- a cup of nuts and bolts

Procedure

1. Obtain a cup of nuts and bolts from your teacher.

2. The nuts and bolts are the reactants. The product consists of 2 nuts on each bolt. Make as many products as possible.

Results/Analysis

1. Using N to symbolize a nut and B to symbolize a bolt, write out an equation for the formation of the product. Pay attention to the difference between a subscript and a coefficient.

2. How many nuts did you have? How many bolts?

3. How many products could you make?

4. Which reactant (nut or bolt) was limiting? How did you make this determination?

5. The limiting reactant was the one that had (fewer/more) pieces.

6. An average mass of a bolt is 10.64 g and an average mass of a nut is 4.35 g. Suppose you are given "about 1500 g" of bolts and "about 1500 g" of nuts. Answer the following questions.
   a. How many bolts are there is "about 1500 g"? How many nuts are there in "about 1500g"?
   b. Which reactant is limiting? How can there be a limiting reactant if you have equal masses of each?
   c. The limiting reactant was the one that had (fewer/more) pieces. Compare this answer to question 5 above. What does it tell you?
   d. What is the largest possible mass of product? How many of the product could you make?
   e. What is the mass of leftover reactant?

Aqueous Solutions

1. Consider Figure 4.2 in the text. Why is it the case that the formulas for ionic compound are always empirical formulas?

2. Label each of the following statements as true or false. Explain your answers and provide an example for each that supports your answer.

   a. All non-electrolytes are insoluble.
   b. All insoluble substances are non-electrolytes.
   c. All strong electrolytes are soluble.
   d. All soluble substances are strong electrolytes.

3. Look at Figure 4.4 in the text. It is possible that a weak electrolyte solution can cause the bulb to glow brighter than a strong electrolyte. Explain how this is possible.

4. When performing calculations that concern dilutions it is best to keep in mind what the solution "looks like" at a molecular level. Answer Active Learning Question 3 at the end of Chapter 4.

5. Using equations to solve problems can be efficient, but it is best if we understand why the equations work. Otherwise these equations may be improperly applied. Explain why the equation $M_1V_1 = M_2V_2$ can be used when solving dilution problems.

6. You have equal masses of different solutes dissolved in equal volumes of solution. Which of the solutes listed below would make the solution with the highest concentration measured in molarity? Defend your answer.

   $NaCl$, $MgSO_4$, LiF, KNO_3

7. Which of the following solutions contains the greatest number of particles? Support your answer.
   a) 400.0 mL of 0.10 *M* sodium chloride
   b) 300.0 mL of 0.10 *M* calcium chloride
   c) 200.0 mL of 0.10 *M* iron(III) chloride
   d) 200.0 mL of 0.10 *M* potassium bromide
   e) 800.0 mL of 0.10 *M* sucrose (table sugar)

8. Do the activity: *Rainbow in a Straw*. You will find the instructions at the end of this chapter.

Precipitation Reactions

1. Consider the mixing of aqueous solutions of lead(II) nitrate and sodium iodide to form a solid.
   a. Name the possible products and determine the formulas of these possible products.
   b. What is the precipitate? How do you know?
   c. Must the subscript for an ion in a reactant stay the same as the subscript of that ion in a product? Explain your answer.

2. What is the purpose of spectator ions? If they are not present as part of the reaction, why are they present at all?

3. Mixing aqueous solutions of sodium chloride and potassium nitrate does not result in a chemical change. Why not?

3. In Section 4.7 of your text, steps for solving stoichiometric problems involving solutions are listed. Step 2 states "Write the balanced net ionic equation for the reaction." Why is it best to write the net ionic equation?

4. As with all quantitative problems in chemistry, make sure not to get "lost in the math". In particular, work on visualizing solutions at a molecular level. For example, consider

   You have two separate beakers with aqueous solutions, one with 4 "units" of potassium sulfate and one with 3 "units" of barium nitrate.

   a. Draw molecular-level diagrams of both solutions.
   b. Draw a molecular-level diagram of the mixture of the two solutions before a reaction has taken place.
   c. Draw a molecular-level diagram of the product and solution formed after the reaction has taken place.

5. After doing problem 4 above, now consider a similar problem with volumes and concentrations.

   You have the following: 2.00 L of a 2.00 M solution of $AgNO_3(aq)$ called solution A and 3.00 L of a 1.00 M solution of $Na_2CrO_4(aq)$ called solution B. Mixing of these solutions makes solution C.

   a. Draw two pictures, one of solution A and one of solution B. Make sure your pictures show the relative volumes of solutions and the relative numbers of ions.
   b. Draw a picture of solution C before any reaction has taken place. Make sure this picture shows the relative volume of solution C and the relative number of ions, as compared with solutions A and B.
   c. Draw a picture of solution C after the reaction has taken place. Make sure this picture shows the relative volume of solution C and the relative number of ions, as compared with solutions A and B.
   d. Calculate the concentrations (in *M*) of all ions in solution C.

Oxidation-Reduction Reactions

1. Make a list of nitrogen compounds with as many different oxidation states for nitrogen as you can.

2. Each of the following results in a chemical reaction. Which of these is not a redox reaction? In all cases, determine the oxidation states for each of the elements all species.
   a. Methane gas is burned in air.
   b. A piece of silver metal is placed in an aqueous solutions of copper(II) nitrate.
   c. Aqueous solutions of lead(II) nitrate and sodium iodide are mixed together.

3. Which of the following must be a redox reaction? Explain your answer and include an example redox reaction for all that apply.
   a. A metal reacts with a nonmetal.
   b. A combustion reaction.
   c. A precipitation reaction.
   d. An acid-base reaction.

4. What does it mean for a substance to be *oxidized*? The term oxidation originally came from substances reacting with oxygen gas. Explain why a substance which reacts with oxygen gas will always be "oxidized".

5. If an element is a reactant or product in a chemical reaction, the reaction must be an oxidation-reduction reaction. Why is this true?

6. The equation $Ag^+ + Cu \rightarrow Cu^{2+} + Ag$ has equal numbers of each type of element on each side of the equation. This equation, however, is not balanced. Why is this equation not balanced? Balance the equation.

Rainbow in a Straw

Materials

- five paper cups containing five different colored solutions on NaCl
- clear, colorless plastic straw

Procedure

1. Obtain five paper cups containing five different colored solutions. You will also need a clear, colorless plastic straw.
2. Each of your five solutions has a different concentration of NaCl. Your task is to order the solutions from least concentrated to most concentrated. You may only use the five solutions and the straw. You may need to experiment a little and make some mistakes along the way.

Results/Analysis

1. How does concentration of the solution relate to density of the solution?
2. List the order of solutions from least concentrated to most concentrated.

Gas Laws

1. Gases are said to exert pressure. Provide a molecular-level explanation for this.

2. How does a barometer measure atmospheric pressure? If the atmospheric pressure increases, how does this register on the barometer and why? What if the atmospheric pressure decreases?

3. The text discusses a mercury barometer. How would a barometer made with water differ from a mercury barometer? Explain and quantify your answer.

4. What is meant by the term "volume" of a gas? Provide a molecular level explanation for this.

5. What is meant by the term "temperature"? What does temperature measure?

6. Do the *Gas Laws and Drinking Straws* activity. You will find the instructions at the end of this chapter.

7. A graph of volume vs. temperature (in Kelvin) is a straight line with a positive slope because volume and temperature (K) are directly related. Pressure and volume are inversely related. Why isn't a graph of pressure vs. volume a straight line with a negative slope? Sketch a graph of pressure vs. volume and justify its shape.

8. Why is it incorrect to say that a sample of helium at 50°C is twice as hot as a sample of helium at 25°C?

9. We can use different units for pressure or volume, but we must use units of Kelvin for temperature. Why must we use the Kelvin temperature scale?

10. Use the ideal gas law equation to derive formulas relating
    a. Pressure and number of moles of ideal gas at constant temperature and volume.
    b. Number of moles of ideal gas and temperature (K) at constant pressure and volume.

    Make a graph of each of the above relationships.

11. On a beautiful autumn day in Champaign, Illinois, you are holding two balloons, each with 15 moles of gas. One is filled with argon gas, and the other is filled with neon gas. Estimate the volume of each balloon.

12. Do the *Rising Water* activity. You will find the instructions at the end of this chapter.

13. You have helium gas in a two-bulbed container connected by a valve as shown below. Initially the valve is closed.

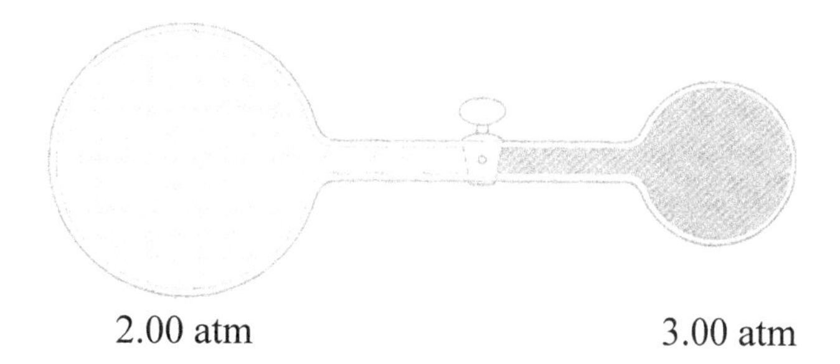

   a. When the valve is opened will the total pressure in the apparatus be less than 5.00 atm, equal to 5.00 atm, or greater than 5.00 atm? Explain your answer.
   b. The left bulb has a volume of 9.00 L and the right bulb has a volume of 3.00 L. Calculate the final pressure after the valve is opened.

Gas Stoichiometry

1. Discuss what happens to the density of a gas as you increase the temperature if the gas is in
   a. a rigid steel container.
   b. a container fitted with a movable piston.

2. Helium floats in air because it is less dense than air. What is the ratio of the densities of air: helium at the same pressure and temperature?

3. If the air temperature is 25°C, to what temperature must you cool a helium balloon so that it no longer floats in the air?

4. You are holding two balloons, an orange balloon and a blue balloon. The orange balloon is filled with neon (Ne) gas, and the blue balloon is filled with argon (Ar) gas. The orange balloon has twice the volume of the blue balloon. Determine the mass ratio of Ne:Ar in the balloons.

5. At a given temperature and pressure, 2.0 L of N_2 gas reacts completely with 6.0 L of H_2 gas to form 4.0 L of product. What is the formula of the product? Support your answer.

6. Estimate the mass of air at normal conditions that takes up the volume of your head. Provide support for your answer.

7. You have a certain mass of helium gas in a rigid steel container. You add the same mass of neon gas to this container. Which of the following best describes what happens? Assume the temperature is constant and support your answer.
   a. The pressure in the container doubles.
   b. The pressure in the container increases but does not double.
   c. The pressure in the container more than doubles.
   d. The volume of the container doubles.
   e. The volume of the container more than doubles.

8. You are holding two balloons of the same volume. One balloon contains 1.0 g helium. The other balloon contains neon. Calculate the mass of neon in the balloon.

9. A common joke on television and in the comics is to have a child grab some helium balloons at a carnival or circus and suddenly be lifted into the sky. But is this possible?
   a. Estimate the minimum number of 10-L helium balloons it would take to lift an average child. Support your answer.
   b. In addition, explain why the balloons can lift the child at all (don't just say "helium is lighter than air" – the explanation should be deeper than that).

Partial Pressures

1. In Section 5.5 of your text, it is stated that the total number of gas particles (in a given volume at a given temperature) determines the pressure of the gas and the identities of the particles are not important. The text goes on to state that this idea tells us two things about gases. State these and explain how Dalton's law supports them.

2. Consider a mixture of equal masses of helium and hydrogen gases. Which gas exerts the greater partial pressure? By what factor? Support your answer.

3. You have two containers with the same volume at the same temperature, each with 2 moles of gas. The first container has a 1:1 mole ratio of neon to argon, and the second has a 1:1 mass ratio of neon to argon.
   a. In which container is the partial pressure of neon greater?
   b. What is the ratio of partial pressure of neon in container #1 to partial pressure of neon in container #2?

The Kinetic Molecular Theory of Gases and Real Gases

1. Use the kinetic molecular theory to explain why we need to use the Kelvin scale when dealing with gases.

2. Visualizing a gas sample at a molecular level is a good way to understand the kinetic molecular theory. Answer problem 38 at the end of Chapter 5.

3. Consider the oxygen and nitrogen molecules composing the air in this room. Determine their average speed in units of miles per hour.

4. At the same temperature, helium atoms have a higher average velocity than oxygen molecules, thus helium atoms have a greater number of collisions with the walls of the container than do the oxygen molecules. If we have an equimolar (equal number of moles) mixture of He and O_2, the partial pressures of each gas are equal. How can this be?

5. Indicate which of the graphs below represents each plot described in questions a-i. Note: the graphs may be used once, more than once, or not at all.

a) 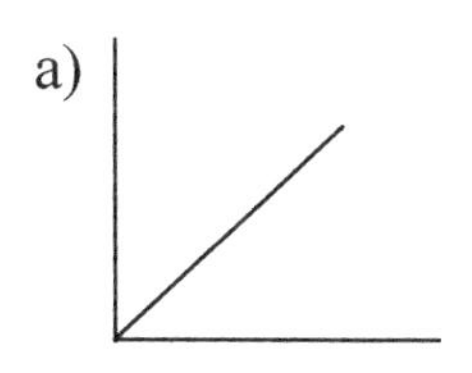

b) 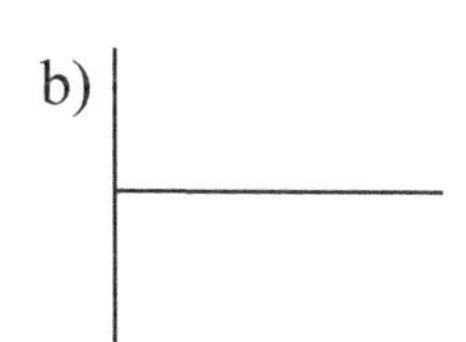

c) 

d) 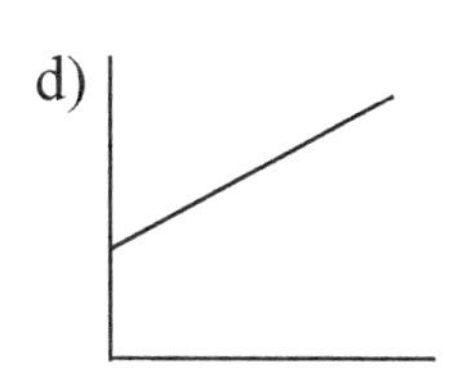

e) 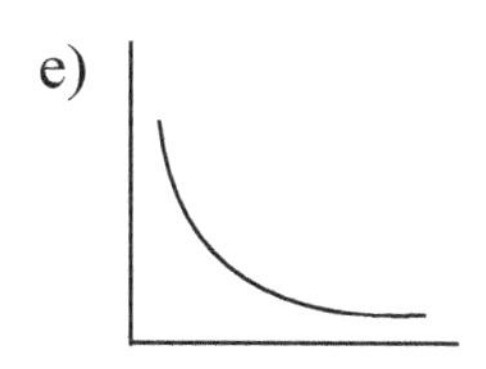

a. n (y) vs. T (x) for an ideal gas at constant P and V.
b. PV (y) vs. V (x) for 1.0 mol of an ideal gas at constant T.
c. Gas density (y) vs. T(x) for 1.0 mol of an ideal gas at constant P.
d. PV (y) vs. n (x) for an ideal gas at constant T.
e. Pressure vs. volume at constant temperature and moles.
f. Volume vs. temperature (°C) at constant pressure and moles.
g. Volume vs. temperature (K) at constant pressure and moles.
h. *PV* vs. *P* at constant temperature for 1 mole of gas.
i. Volume vs. moles at constant temperature and pressure.

6. Explain each of the following scenarios (a-c). You are to provide both the law (in words and a formula) and the theory.

A complete answer will do the following:

- Use the ideas of the kinetic molecular theory (talk about atoms/molecules) and use the terms "pressure", "temperature", "volume", and "moles of gas" in your answer.
- Discuss which factors change and which are constant, how you know this, and how this affects what happens.

a. You pack a lunch in your house on the beach, and head up to the mountains. When you get to the top, you unpack your lunch and notice your sealed bag of potato chips has expanded. Why has this occurred? The temperature on the beach and in the mountains is approximately the same.
b. You are eating a late night snack and look at the can of aerosol cheese in your hand. There is a warning label which reads: "Warning: Contents under pressure. Do not heat". Your roommate says "Cool. Let's heat it and see what happens." Explain to your roommate what would happen and why (by the way, don't try this!).
c. You and a friend are sailing in a hot-air balloon. The balloon (which has a large opening at the bottom and a smaller one at the top) is fully expanded as you fly overhead. You turn on the burner at the bottom opening of the balloon to increase the heat and the balloon goes up even further. Explain how this works.

7. Is the observed pressure of a real gas greater than, less than, or equal to the ideal pressure? Explain why this is true.

8. Is the volume of the vessel containing a gas greater than, less than, or equal to the volume available to the real gas particles in the sample.

Gas Laws and Drinking Straws

Materials

- Two drinking straws
- Glass or cup

Procedure 1

1. Half fill a cup with water.
2. Place the ends of two drinking straws in your mouth.
3. Submerge the other end of one straw into the water, and leave the other straw out of the water (on the side of the cup).
4. Try to drink the water. What do you notice?

Procedure 2

1. Place a drinking straw vertically into a cup that is half-filled with water.
2. Place your finger over the opening of the straw and take the straw out of the water. What happens? Make careful observations.

Results/Analysis

Explain your observations for each of the two procedures.

Rising Water

Materials

- candle
- matches
- glass
- water
- test tube
- 400 mL beaker
- 150 mL beaker
- Food coloring
- Hot plate

Procedure 1

1. Light a candle and let some wax drip onto the bottom of a glass which is taller than the candle. Blow out the candle and fix the candle to the glass before the wax solidifies.
2. Add water to the glass until the candle is about half submerged. Do not get the wick of the candle wet.
3. Light the candle. Take a test tube which is wider than the candle and quickly place it over the candle and submerge the opening of the test tube under the water. What happens? Make careful observations.

Procedure 2

1. Fill the 400 mL beaker with about 200 mL water and add a couple drops of food coloring.
2. Place the 150 mL beaker upside-down in the 400 mL beaker.
3. Heat the water to boiling.
4. After the water has boiled for a few minutes, remove the beakers from the heat. What happens? Make observations.

Results/Analysis

Explain your observations for each of the two procedures.

The Nature of Energy

1. Look at Figure 6.1 in your text. Ball A has stopped moving. However, energy must be conserved. So what happened to the energy of Ball A?

2. What if energy was not conserved? How would this affect our lives?

3. The text uses distance traveled and change in elevation to discuss the idea of a state function. Explain which of these is a state function and which is not.

4. A friend of yours reads that the process of water freezing is exothermic. This friend tells you that this can't be true because exothermic implies "hot", and ice is cold. Is the process of water freezing exothermic? If so, explain it so your friend can understand it. If not, explain why not.

5. Label the following processes as exothermic or endothermic and explain:
   a. Your hand gets cold when you touch ice.
   b. The ice gets warmer when you touch it.
   c. Water boils in a kettle upon being heated on a stove.
   d. Water vapor condenses on a cold pipe.
   e. Ice cream melts.

6. The internal energy of a system is said to be the sum of the kinetic and potential energies of all the particles in the system. The text discusses *potential energy* and *kinetic energy* in terms of a ball on a hill in Section 6.1. Explain *potential energy* and *kinetic energy* for a chemical reaction.

7. You strike an unlit match and it burns.

   Explain the energy transfers of this scenario using the terms exothermic, endothermic, system, surrounding, potential energy, and kinetic energy in your discussion. Also include an energy-level diagram.

8. For each of the following, define a system and its surroundings, and give the direction of energy transfer.
   a. Propane is burning in a Bunsen burner in a laboratory.
   b. Water drops, sitting on your skin after swimming, evaporate.
   c. Two chemicals mixing in a beaker give off heat.

9. Hydrogen gas and oxygen gas react violently to form water.
   a. Which is lower in energy: a mixture of hydrogen gases, or water? Explain.
   b. Sketch an energy-level diagram (like Figure 6.3 or Figure 6.4) for this reaction.

10. Which of the following performs more work?

    A gas expanding against a pressure of 2 atm from 1.0 L to 4.0 L

    or

    A gas expanding against a pressure of 3 atm from 1.0 L to 3.0 L

11. Determine the sign of ΔE for each of the following with the listed conditions:
    a. An endothermic process that performs work.
        i. $|\text{work}| > |\text{heat}|$
        ii. $|\text{work}| < |\text{heat}|$
    b. Work is done on a gas and the process is exothermic.
        i. $|\text{work}| > |\text{heat}|$
        ii. $|\text{work}| < |\text{heat}|$

12. There is a law of conservation of energy (the first law of thermodynamics). Is there a law of conservation of heat? Defend your answer.

Enthalpy and Calorimetry

1. Consider four 100.0-g samples of water, each in a separate beaker at 25.0°C. Into each beaker you drop 10.0 g of a different metal that has been heated to 95.0°C. Assuming no heat loss to the surroundings, which water sample will have the highest final temperature? Explain your answer.
    a) The water to which you have added aluminum (c=0.89J/g°C).
    b) The water to which you have added iron (c=0.45J/g°C).
    c) The water to which you have added copper (c = 0.20J/g°C).
    d) The water to which you have added lead (c=0.14J/g°C).
    e) Since the masses of the metals are the same, the final temperatures would be the same.

2. Explain why aluminum cans make good storage containers for soft drinks.

3. A 100.0 gram sample of water at 90°C is added to a 100.0-gram sample of water at 10°C.
    a. The final temperature of the water should be
        i. Between 50°C and 90°C
        ii. 50°C
        iii. Between 10°C and 50°C
    b. Calculate the final temperature of the water.

4. A 100.0-gram sample of water at 90°C is added to a 500.0-gram sample of water at 10°C.
    a. The final temperature of the water should be
        i. Between 50°C and 90°C
        ii. 50°C
        iii. Between 10°C and 50°C
    b. Calculate the final temperature of the water.

5. You have a Styrofoam cup with 50.0 g of water at 10°C. You add a 50.0-gram iron ball at 90°C to the water.
    a. The final temperature of the water should be
        i. Between 50°C and 90°C
        ii. 50°C
        iii. Between 10°C and 50°C
    b. Calculate the final temperature of the water.

Hess's Law and Standard Enthalpies of Formation

1. How is Hess's law a restatement of the first law of thermodynamics?

2. In Section 6.3 of your text, two characteristics of enthalpy changes for reactions are listed. What are these characteristics? Explain why these characteristics are true.

3. Look at equation 6.1 in Section 6.4 of your text. Does this mean $\Delta H°_{reaction}$ is or is not a state function? Explain your answer.

4. Explain why $\Delta H°_f$ for an element in its standard state is zero.

5. Using the following data, calculate the standard heat of formation of the compound $ICl(g)$ at 25°C, and show your work.

	$\Delta H°$ (kJ/mol)
$Cl_2(g) \rightarrow 2Cl(g)$	242.3
$I_2(g) \rightarrow 2I(g)$	151.0
$ICl(g) \rightarrow I(g) + Cl(g)$	211.3
$I_2(s) \rightarrow I_2(g)$	62.8

The Nature of Matter and The Atomic Spectrum of Hydrogen

1. Consider Figure 7.1 in your text. Determine the relative wavelengths and energies for the three waves.

2. Explain what is meant by the term "excited state" as it applies to an electron. Is an electron in an excited state higher or lower in energy than an electron in the ground state? Is an electron in an excited state more or less stable than an electron in the ground state?

3. What does it mean when we say energy levels are *quantized*?

4. What evidence do we have that energy levels in an atom are quantized? State and explain the evidence.

5. Explain the hydrogen emission spectrum. Why is it significant that the color emitted is not white? How does the emission spectrum support the idea of quantized energy levels?

6. There are an infinite number of allowed transitions in the hydrogen atom. Why don't we see more lines in the emission spectrum for hydrogen?

The Bohr Model and the Quantum Mechanical Model of the Atom

1. Note the equation we use to calculate the change in energy associated with an electron in the hydrogen atom: $E = -2.178 \times 10^{-18}\ J(Z^2/n^2)$

   a. Why is this energy "negative"?
   b. What does "zero energy" represent/mean?
   c. Why does it make sense that as Z increases, E becomes more negative (what is Z, by the way)?
   d. Why does it make sense that as n increases, E becomes less negative (what is n, by the way)?

2. Describe what is meant by the term *orbital*. Include a discussion of probability.

3. Explain the difference between *probability distribution* and *radial probability*. See Figures 7.12 and 7.13 in your text and use these in your discussion.

4. Do the *Electron Probability* activity. You will find the instructions at the end of this chapter.

Quantum Numbers and Orbital Shapes and Energies

1. One way of demonstrating you have an understanding of applying rules is to change the rules slightly and see how this affects the results. Problem 181 at the end of Chapter 7 in your text changes the rules of determining quantum numbers. Answer this question.

2. Figure 7.14 in your text shows 1*s*, 2*s*, and 3*s* orbitals for the hydrogen atom.
   a. How are the orbitals similar?
   b. How are the orbitals different?
   c. How can a hydrogen atom have a 2*s* orbital if it has only one electron?

3. If energy is added to a hydrogen atom such that an electron has enough energy to "jump" to the n=2 level, is the electron more likely to be in a 2s orbital or 2p orbital? Explain your answer, and use Figure 7.19 in your text in your explanation.

4. In summarizing the hydrogen atom, your text states that the size of an orbital is arbitrarily defined. How is it defined, and why must the size of an orbital be arbitrarily defined?

Polyelectronic Atoms

1. Compare Figure 7.19 to Figure 7.23 and discuss what these figures tell you. Use Figures 7.21 and 7.22 in your explanation.

2. An electron in the 2p orbital is on average closer to the nucleus than an electron in the 2s orbital. Yet, for elements with greater than 2 electrons (Li and on) the electron chooses to be on the 2s orbital, why is this so?

3. Explain what is meant by the "penetration effect" with respect to orbitals, and discuss two distinct observations it helps to explain.

4. Show how you can use the periodic table to determine the order in which orbitals fill in polyelectronic atoms (so that you do not have to memorize it).

Periodic Trends in Atomic Properties

1. Which atom is larger, Na or Cl? Why?

2. Which atom is larger, Li or Cs? Why?

3. Which atom requires more energy to remove an electron, Na or Cl? Why?

4. Which atom requires more energy to remove an electron, Li or Cs? Why?

5. In going across a row of the periodic table, protons and electrons are being added and atomic radius generally decreases (fluorine has a smaller radius than lithium, for example). In going down a column of the periodic table, protons and electrons are also being added, but the atomic radius generally increases (iodine is larger than fluorine, for example). Explain why this is true.

6. The first ionization energy for a given atom in Group 2 is "x". A good estimate for the second ionization energy of this atom is (defend your answer)
   a) less than "x". It is easier to remove the second electron since this gives the species a noble gas electron configuration.

b) about “2x”. The second electron is harder to remove than the first electron.
c) about “x”. Since the electrons are taken from the same energy level, the ionization energies are about the same.
d) about “-x”. The ionization energy is exothermic since the Group 2 atoms want to lose two electrons to achieve a noble gas electron configuration.
e) about “-2x”. The ionization energy is very exothermic since the Group 2 atoms want to lose two electrons to achieve a noble gas electron configuration.

7. Atom A has valence electrons that are lower in energy than the valence electrons of Atom B. Which atom has the higher ionization energy? Explain.

8. Which is larger, the hydrogen 1*s* orbital or the Li 1*s* orbital? Why? Which is lower in energy, the hydrogen 1*s* orbital or the Li 1*s* orbital? Why?

Electron Probability

According to modern atomic theory, we cannot be sure of the exact location of electrons in an atom. We predict that electrons will be relatively close to the nucleus (because the electrons are negatively charged and the nucleus is positively charged). However, we discuss the "location" of an electron in terms or probability instead of an exact position.

In this activity you will construct and analyze a probability map using a dart and dartboard.

Materials

Goggles
Apron
Darts
Graph paper
Target
Cardboard

Safety

Be careful to drop the darts toward the target on the floor. Nobody should ever be in the path of a dart.

Procedure

1. Tape the target to the center of the cardboard, place it on the floor, and tape this to the floor.
2. Drop the dart from shoulder height trying to hit the center of the target. Your partner should retrieve the dart and mark the position of the hit with a small x (don't count drops that fall outside the largest circle).
3. Repeat this procedure 99 times for a total of 100 drops.
4. Count the number of hits in each ring and record this number.

Data/Observations

Fill in the following table.

Ring Number	Average distance from target center (cm)	Area of ring (cm^2)	Number of hits in the ring	Number of hits per unit area (hits/cm^2)
1	0.5	3.1		
2	1.5	9.4		
3	2.5	16		
4	3.5	22		
5	4.5	28		
6	5.5	35		
7	6.5	41		
8	7.5	47		
9	8.5	53		
10	9.5	60		

Analysis and Conclusions

1. Which is the ring with the highest probability of finding a hit?

2. Which is the ring with lowest probability of finding a hit?

3. Construct a graph of the number of hits vs. average distance from the center.

4. Construct a graph of hits per unit area vs. average distance from the center.

5. What does the graph of the number of hits vs. average distance represent? Account for its shape.

6. What does the graph of the hits per unit area vs. average distance represent? Account for its shape.

7. Is the maximum of each graph the same? Explain.

8. How well can we model an orbital with a dartboard and a dart? Specifically, focus on the following:
   a. Compare your target with Figure 7.11 in your text. How is it similar? How is it different?
   b. Why do we predict an electron should be near the nucleus? Why do we expect the dart to land in the center of the target?

Chemical Bonds and Electronegativity

1. Consider Figure 8.1 in your text. Describe in your own words what this figure tells you. Make sure to discuss what is meant by zero energy, negative energy, and positive energy.

2. What is meant by the term *chemical bond*?

3. Why do atoms bond with each other to form molecules? How do atoms bond with each other to form molecules?

4. Explain the difference between ionic bonding and covalent bonding. How can we use the periodic table to help us determine the type of bonding between atoms?

5. If lithium and fluorine react, which has more attraction for an electron? Why?

6. In a bond between fluorine and iodine, which has more attraction for an electron? Why?

7. Explain the term *electronegativity*, in your own words. What are the trends across and down the periodic table for electronegativity? Explain these trends.

8. True or false? In general, a smaller atom has a larger electronegativity. Explain.

9. Explain how you can use the periodic table to predict relative polarity of bonds. For example, how do you know by looking at the periodic table that a C-O bond is more polar than a N-O bond?

10. We can use the periodic table to predict that a bond between phosphorus and fluorine will be polar, but we cannot make such a prediction for a bond between carbon and sulfur. Explain.

11. Arrange the following bonds from most to least polar and indicate the bond polarity for each (show the partial positive and partial negative charges).

    a. N-F   O-F   C-F
    b. C-F   N-O   Si-F
    c. H-Cl   B-Cl   S-Cl

12. Which of the following bonds would be the least polar yet still be considered polar covalent? Explain.

    Mg-O   C-O   O-O   Si-O   N-O

13. Which of the following bonds would be the most polar without being considered ionic? Explain.

    Mg-O   C-O   O-O   Si-O   N-O

Ions, Ionic Compounds, and Bond Energies

1. Explain how you can use the periodic table to predict the formula of compounds.

2. Choose an alkali metal, an alkaline metal, a noble gas, and a halogen so that they constitute an isoelectronic series when the metals and halogen are written as their most stable ions.

   a. What is the electron configuration for each species?
   b. Determine the number of electrons for each species.
   c. Determine the number of protons for each species.
   d. Rank the species according to increasing radius.
   e. Rank the species according to increasing ionization energy.

3. In Section 8.4 your text states "In virtually every case the atoms in a stable compound have a noble gas arrangement of electrons." While this tells us *what* happens, it doesn't tell us *why*. However, sometimes students use a phrase similar to "atoms want to have a noble gas electron configuration." This statement implies that a noble gas electron configuration is more stable than the electron configuration for a free atom. Use Figure 8.11 in your text to show that the sodium atom, for example, is more stable than the Na^+ ion, and that the oxygen atom is more stable than the O^{2-} ion. Given that the ions are less stable than the atoms, why do the ionic compounds form?

4. The reaction of hydrogen gas reacting with oxygen gas to form water is very exothermic. Use this information to predict which bonds are stronger, the bonds of H_2 and O_2 or the bonds in H_2O. Use the data in Table 8.5 to verify your answer.

Lewis Structures

1. Why do we only consider the valence electrons in drawing Lewis structures?

2. Does a Lewis structure indicate which electrons came from which atoms? Explain.

3. How do we determine the total number of valence electrons for an ion? Provide an example of an anion and a cation and explain your answer.

4. When we draw Lewis structures for polyatomic anions, can we say to which atom the added electron goes? Explain.

5. Draw two different Lewis structures for molecules with the formula C_2H_6O.

6. When we exceed the octet rule, where do the "extra electrons" go? If more than one atom can break the octet rule in a molecule, what do we do?

7. Draw all valid Lewis structures for the molecule N_2O (NNO not NON) and rank these based on formal charge.

Molecular Structure: The VSEPR Model

1. Do the *Geometric Balloons* activity. You will find the instructions at the end of this chapter.

2. What is the main idea in the Valence Shell Electron Pair Repulsion theory?

3. Why must a Lewis structure for a molecule be drawn before we can determine its molecular structure?

4. The molecules NH_3 and BF_3 have the same general formula (AB_3) but different shapes.
   a. Find the shape of each of the above molecules.
   b. Provide more examples of real molecules that have the same general formulas but different shapes.

5. How do we deal with multiple bonds with VSEPR theory?

6. In Section 8.13 of your text the term "effective pairs" is used. What does this mean?

7. Provide two examples of molecules which have a bent or V-shape, but have different bond angles.

8. True or false: Lone pairs make a molecule polar. If true, explain why. If false, provide a counterexample.

Geometric Balloons

Materials

- Twenty round balloons

Procedure

1. Obtain twenty round balloons.
2. Blow up all of the balloons to approximately the same size and tie each one.
3. Make 5 different geometries by: tying two of the balloons together, three of the balloons together, four of the balloons together, five of the balloons together, and six of the balloons together.
4. Observe the geometry of each of the balloon figures. What are the angles between each balloon for each balloon figure?

Results/Analysis

1. Relate your findings to VSEPR theory. What does the knot in the center of each cluster represent? What does each balloon represent? Why do these geometries occur naturally?

2. Provide an example of a real molecule for each of the 5 geometries.

3. Consider the cluster with three balloons.
   a. What is the arrangement of atoms (shape) if one balloon represents a lone pair of electrons?
   b. Provide an example of a real molecule for part a.

4. Consider the cluster with four balloons.
   a. What is the arrangement of atoms (shape) if one balloon represents a lone pair of electrons?
   b. What if there were two lone pairs of electrons?
   c. What if there were three lone pairs of electrons?
   d. Provide an example of a real molecule for each of the above.

5. Consider the cluster with five balloons.
   a. What is the arrangement of atoms (shape) if one balloon represents a lone pair of electrons?
   b. What if there were two lone pairs of electrons?
   c. What if there were three lone pairs of electrons?
   d. Provide an example of a real molecule for each of the above.

6. Consider the cluster with six balloons.
   a. What is the arrangement of atoms (shape) if one balloon represents a lone pair of electrons?
   b. What if there were two lone pairs of electrons?
   c. Provide an example of a real molecule for each of the above.

Hybridization and the Localized Electron Model

1. Do the *Hybridization* activity. You will find the instructions at the end of this chapter.

2. How are molecular geometries related to the hybridization of the central atom? Provide an example of an actual molecule for each geometry.

The Molecular Orbital Model

1. Explain the basic ideas of the molecular orbital model. How does a molecular orbital differ from an atomic orbital? How does a molecular orbital differ from a hybridized orbital?

2. How do the shapes of the molecular orbitals show you whether the orbital is a bonding MO or an antibonding MO? Explain.

3. How does the molecular orbital model account for the stability of the H_2^- ion? Why is the localized electron model a poor model in describing H_2^-?

4. Bond energy has been defined in the text as the amount of energy required to break a chemical bond, so we have come to think of the addition of energy as breaking bonds. However, in some cases, the addition of energy can cause the formation of bonds. For example, in a sample of helium gas subjected to a high energy source, some He_2 molecules exist momentarily and then dissociate. Use MO theory (and diagrams) to explain why the He_2 molecules may come to exist from excited helium atoms, and why they dissociate.

5. Explain the difference in shapes between the σ and π molecular orbitals. Consider the following questions as well.
   a. Why are there two π MOs and one σ MO?
   b. Why are the π MOs degenerate?
   c. Explain the difference in shapes between the bonding and antibonding MOs (for both σ and π).

6. Consider Figure 9.38 in your text.
   a. In general, how does bond order relate to bond dissociation energy? Why is this true?
   b. In general, how does bond order related to bond length. Why is this true?

7. For simple homonuclear molecules, how does bond order relate to the type of bond seen in a Lewis structure (see the molecules in Figure 9.38 of your text, for example).

8. Heteronuclear diatomic molecules require you to consider the difference in the energies of the atoms involved. Problem 61 in your text is a good problem to consider. Do this problem.

9. Why is the concept of resonance not needed if we consider delocalized electrons?

Hybridization

Materials

- Toothpicks
- Clay (different colors if possible)

Procedure and Analysis

1. Draw the Lewis structure for the methane molecule (CH_4) and make a model of it with clay and toothpicks. Use VSEPR theory to construct the correct shape of the molecule.
   a. What are the bond angles in methane? What is the shape of the molecule?
   b. Why can't the bonding orbitals for methane be formed by an overlap of the atomic orbitals (that is, the *s* and *p* orbitals)? What would the molecule look like if this were true? Make an additional model to show this.
   c. Why are sp^3 orbitals called sp^3 orbitals? That is, why *s*, why *p*, why 3?
   d. Draw Lewis structures and make models for NH_3 and H_2O. Use these drawings and models to show how the central atoms for CH_4, NH_3, and H_2O are sp^3 hybridized.

2. Draw the Lewis structure for the ethene molecule (C_2H_4) and make a model of it with clay and toothpicks. Use VSEPR theory to construct the correct shape of the molecule.
   a. What are the bond angles in ethene? What is the shape of the molecule?
   b. Why can't sp^3 hybridization account for the ethene molecule? Explain why the carbons in ethane must be sp^2 hybridized.
   c. In sp^2 hybridization, 2 of the *p* orbitals for each carbon atom are being used for the hybrid orbital. What happens to the third *p* orbital?
   d. Make a model of ethene to show the unhybridized *p* orbital on each carbon atom. Consider Figure 9.13 in your text and make sense of it. Label the bonds as σ or π.

3. Draw the Lewis structure for the carbon dioxide molecule (CO_2) and make a model of it with clay and toothpicks. Use VSEPR theory to construct the correct shape of the molecule.
   a. What are the bond angles in carbon dioxide? What is the shape of the molecule?
   b. What is the hybridization of the carbon atom in carbon dioxide? Support your answer.
   c. What is the hybridization of each oxygen atom in carbon dioxide? Support your answer.
   d. Make a model of carbon dioxide to show the hybrid and unhybridized orbitals (like in Figure 9.19 in your text). Label the bonds as σ or π.

4. Draw the Lewis structure for the hydrogen cyanide molecule (HCN) and make a model of it with clay and toothpicks. Use VSEPR theory to construct the correct shape of the molecule.
   a. What are the bond angles in hydrogen cyanide? What is the shape of the molecule?
   b. What is the hybridization of the carbon atom in hydrogen cyanide? Support your answer.
   b. What is the hybridization of the nitrogen atom in hydrogen cyanide? Support your answer.
   d. Make a model of hydrogen cyanide to show the hybrid and unhybridized orbitals (like in Figure 9.19 in your text). Label the bonds as σ or π.

Intermolecular Forces and Physical Properties

1. Sketch a microscopic picture of water and distinguish between *intramolecular bonds* and *intermolecular forces*. Which correspond to the bonds we draw in Lewis structures?

2. Which has the stronger intermolecular forces, N_2 or H_2O? Explain.

3. Which gas would behave more ideally at the same conditions of pressure and temperature, CO or N_2? Why?

4. Wax consists of long chain non-polar hydrocarbons that exhibit London dispersion forces. Water exhibits hydrogen bonding. Individual hydrogen bonding interactions are stronger than individual London dispersion interactions. So how can wax be a solid and water a liquid at normal conditions?

5. Explain the relative freezing points as seen in Table 10.2. Discuss the nature of London dispersion forces in your answer.

6. True or false? All molecules with polar bonds are polar. If true, explain why and include an example. If false, provide a counterexample and explain it.

7. True or false? Methane (CH_4) is more likely to form stronger hydrogen bonding than water because each methane molecule has twice as many hydrogen atoms. Provide a concise explanation of hydrogen bonding to go with your answer.

8. Why should it make sense that N_2 exists as a gas? Given your answer to this, how is it possible to make liquid nitrogen? Explain why lowering the temperature works.

Structures and Types of Solids

1. The unit cell in this two-dimensional crystal contains __________ Xs and __________ Os.

```
X     X     X     X
   O     O     O     O
X     X     X     X
   O     O     O     O
X     X     X     X
   O     O     O     O
```

2. Determine the number of metal atoms in a unit cell if the packing is
   a. simple cubic
   b. cubic closest packing

3. A metal crystallizes in a face-centered cubic structure. Determine the relationship between the radius of the metal atom and the length of an edge of the unit cell.

4. Explain the difference among ionic, atomic, and molecular solids. Provide an example of each.

5. White phosphorus and sulfur are each called molecular solids even though each is made of only phosphorus and sulfur, respectively. How can they be considered molecular solids? If this is true, why isn't diamond (which is made up only of carbon) a molecular solid?

6. A certain metal chloride crystallizes in such a way that the fluoride ions occupy simple cubic lattice sites, while the metal atoms occupy the body centers of half the cubes.

   List some possibilities for the metal ion.

7. Why is it incorrect to use the term "molecule of NaCl" but correct to use the term "molecule of H_2O"? Is the term "molecule of diamond" correct? Explain.

8. The unit cell in a certain lattice consists of a cube formed by an anion at each corner, an anion in the center, and a cation at the center of each face. Provide two examples of ionic compounds that fit this description.

Phase Changes and Phase Diagrams

1. What is the vapor pressure of water at 100°C? How do you know?

2. Claiming that the boiling point of water is 100°C is not necessarily incorrect, but it is incomplete. Explain why.

3. As intermolecular forces increase, what happens to each of the following? Why?
   a. Boiling point
   b. Viscosity
   c. Surface tension
   d. Enthalpy of fusion
   e. Freezing point
   f. Vapor pressure
   g. Heat of vaporization

4. The boiling point of X is less than that of Y, which is less than that of Z. How do the vapor pressures of X, Y, and Z compare? Explain.

5. Of 1.0 mol samples of the following gases at STP (0°C, 1 atm), which behaves most ideally and which behaves least ideally?

   He, N_2, CO, SO_2

6. At room conditions of pressure and temperature, methane (CH_4) is a gas, water (H_2O) is a liquid, and iodine (I_2) is a solid. Use your understanding of Lewis structures, VSEPR, and intermolecular forces (IMFs) to explain the different phases of these three substances at the same conditions.

7. Which would you predict should be larger for a given substance: ΔH_{vap} or ΔH_{fus}? Explain why.

8. Draw two Lewis structures from the formula C_2H_6O and compare the boiling points of the two molecules.

9. Given below are the temperatures at which two different liquid compounds with the same empirical formula have a vapor pressure of 400 torr.

Compound	T (° C)
A	–37.8
B	63.5

Which of the following statements (a–d) is false? Defend your answer.

   a) Increasing the temperature will increase the vapor pressure of both liquids.
   b) Intermolecular attractive forces are stronger in liquid compound B than in liquid compound A.
   c) The normal boiling point of compound A will be higher than the normal boiling point of compound B.
   d) At the same temperature, compound A would have a higher vapor pressure than compound B.
   e) None of these is false.

10. Explain the differences between the phase diagrams of water and carbon dioxide.

Concentration of Solutions

1. You have an open beaker containing a solution of table salt dissolved in water. What happens to the salt concentration (increases, decreases, or stays the same) over time? Draw molecular level pictures to explain your answer.

2. Must a dilute solution contain less solute than a concentrated solution? Support your answer.

3. Which solution has the higher concentration in terms of molarity: a 2.0 N HCl solution or a 3.0 *N* H_2SO_4 solution? Justify your answer.

The Energetics of Solutions and Solubility

1. Refer to Figure 11.1 when answering the following questions. In your explanations, estimate the ΔH values of the various processes for solution formation (large or small, positive or negative), order them according to magnitude (absolute values), and explain.

   a. Explain why water and oil (a long-chain hydrocarbon) do not mix.
   b. Explain why two different oils (oils are long-chain hydrocarbons) do mix.
   c. Explain why water and another polar liquid do mix.

2. Explain the phrase "like dissolves like". Why does it generally hold true? Provide a molecular level picture to support your answer.

3. Why are carbonated beverages stored in bottles with a relatively high pressure of CO_2 above the solution?

4. In general how does temperature affect the solubility of solids in water? What about gases in water?

The Vapor Pressures of Solutions

1. Would you expect each of the following solutions to be relatively ideal (with respect to Raoult's law), to show a positive deviation, or to show a negative deviation? Explain.
   a. hexane (C_6H_{14}) and chloroform ($CHCl_3$)
   b. ethyl alcohol (C_2H_5OH) and water
   c. hexane (C_6H_{14}) and octane (C_8H_{18})

2. Consider an ideal mixture of two volatile liquids, A and B. Answer the following questions. Use pictures in your explanations.

   a. Suppose you have an equimolar mixture of A and B, and the vapor pressure of pure liquid A is twice that of pure liquid B. Determine the mole fraction of A in the vapor above the mixture.
   b. Suppose you make a mixture with twice as much A as B (in moles), and the vapor pressure of pure liquid A is twice that of pure liquid B. Determine the mole fraction of A in the vapor above the mixture.

c. Suppose you make a mixture with twice as much B as A (in moles), and the vapor pressure of pure liquid A is twice that of pure liquid B. Determine the mole fraction of A in the vapor above the mixture.

3. Benzene and toluene form ideal solutions. Imagine mixing benzene and toluene at a temperature where the vapor pressure of pure benzene is 750.0 torr and the vapor pressure of toluene is 300.0 torr.

   a. You make a solution in which the mole fraction of the benzene is 0.500. You next place this solution in a closed container and wait for the vapor to come into equilibrium with the solution. Then you condense the vapor. Determine the composition (mole percent) of the vapor. Explain why the relative numbers make sense.
   b. You make a solution by pouring some toluene into some benzene. You next place this solution in a closed container and wait for the vapor to come into equilibrium with the solution. Then you condense the vapor and determine the mole fraction of the toluene to be 0.500. Determine the composition (mole percent) of the original solution. Explain why the relative numbers make sense.

4. Would a solution with a positive deviation from Raoult's law have a higher or lower boiling point than ideal? Explain.

5. Consider Figure 11.10 in your text. Over time, the volume of pure water in the one beaker went to zero, and the volume increased in the solution.

   a. Explain this phenomenon using terms such as *vapor pressure* and *equilibrium*. Use these terms appropriately and correctly.
   b. Also, sketch a graph of vapor pressure vs. time for the pure water and aqueous solution (both on the same graph) to help explain what is happening.

Colligative Properties

1. Explain what is meant by the term *colligative properties*.

2. Explain how adding a non-volatile solute increases the boiling point and decreases the freezing point of the solvent.

3. Why is molality (not molarity) used for colligative properties?

4. Explain what is meant by the term *osmotic pressure*.

5. Consider two solutions. In solution A, you add 1.0 mole of table sugar to 1.0 L of water. In solution B, you add 1.0 mole of table salt to 1.0 L of water. Assume ideal behavior.

   a. How does the freezing point of each solution compare to that of water? Explain your answer.
   b. How do the freezing points of these solutions compare to each other? Explain your answer.

Reaction Rates and Rate Laws

1. How do relative reaction rates for reactants and products in a given reaction compare to the mole ratio in a balanced equation? Why is this true?

2. Explain the *method of initial rates* when trying to determine the form of a rate law. Why do we need to determine the *initial* rates?

3. What is the difference between a differential and an integrated rate law? What does each measure?

4. For a reaction with two reactants, why do we make the concentration of one reactant very large with respect to the concentration of the other reactant? How do we determine the rate constant, *k*, for such a reaction?

5. For a reaction aA → Products, make a sketch of a graph of [A] vs. time for zero-, first-, and second-order reactions. Using these graphs, explain the behavior of successive half-lives for reactions with these orders.

6. Consider the reaction aA → Products. $[A]_0 = 5.0\ M$ and $k = 1.0 \times 10^{-2}$. Calculate [A] after 30.0 seconds have passed for each of the following conditions.
   a. The reaction is zero-order.
   b. The reaction is first-order.
   c. The reaction is second-order.

Reaction Mechanisms

1. Your text defines an elementary step as a *reaction whose rate can be written from its molecularity*. What does this mean?

2. How do exponents (orders) in rate laws compare to coefficients in balanced equations? Why?

3. The reaction 2A + B → C has the following proposed mechanism:

   $A + B \rightleftharpoons D$ (fast equil.)
   $D + B \rightarrow C$ (slow)

   Write the rate law from this mechanism.

4. The experimental rate law for the decomposition of nitrous oxide (N_2O) to N_2 and O_2 is rate = $k[N_2O]^2$. Which of the following could be a correct mechanism? Defend your answer.

   I. $N_2O \rightarrow N_2 + O$
   $N_2O + O \rightarrow N_2 + O_2$

II. $2N_2O \rightleftharpoons N_4O_2$
$N_4O_2 \rightarrow 2N_2 + O_2$

a) Mechanism I with the first step as the rate-determining step.
b) Mechanism I with the second step as the rate-determining step as long as the first step is a fast equilibrium step.
c) Mechanism II with the second step as the rate-determining step if the first step is a fast equilibrium step.

5. You are studying the kinetics of the reaction $H_2(g) + F_2(g) \rightarrow 2HF$ and you wish to determine a mechanism for the reaction.

You run the reaction twice by keeping one reactant at a much higher pressure than the other reactant (this lower pressure reactant begins at 1.00 atm). Unfortunately, you neglect to record which reactant was at the higher pressure, and you forget which it was later. Your data for the first experiment is:

Pressure of HF (atm)	time (min)
0	0
0.3000	30.00
0.6000	65.84
0.9000	110.4
1.200	169.1
1.500	255.9

When you ran the second experiment (in which the higher pressure reactant was run at a much higher pressure), you determine the values of the apparent rate constants to be the same.

It also turns out that you find data taken from another person in the lab. This individual found that the reaction proceeds about 40 times as fast at 55°C as at 35°C.

You also know, from the energy level diagram, that there are three steps to the mechanism, and the first step has the highest activation energy. You look up the bond energies of the species involved and they are (in kJ/mol): H-H (432), F-F (154), H-F (565).

a. Develop a reasonable mechanism for the reaction. Support your answer and explain the significance of each piece of information above.
b. Which reactant was limiting in the experiments?

6-8. Choose the correct graph for the plots described below.

a) 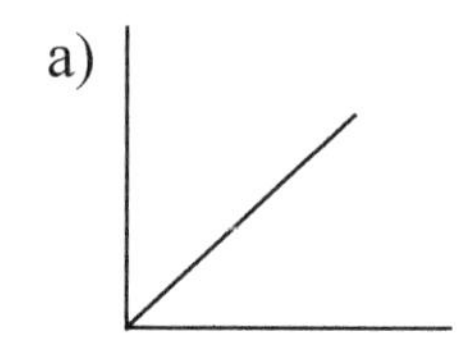

b) 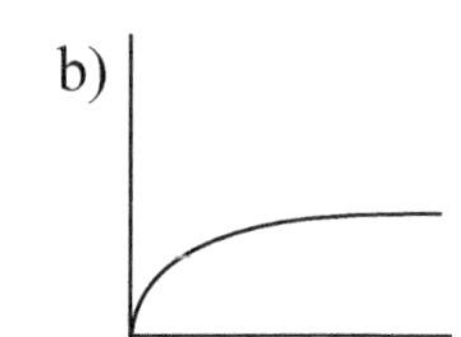

c) 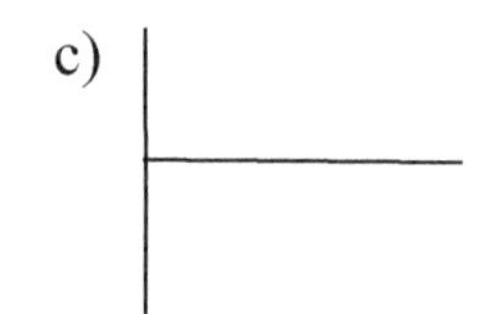

d) 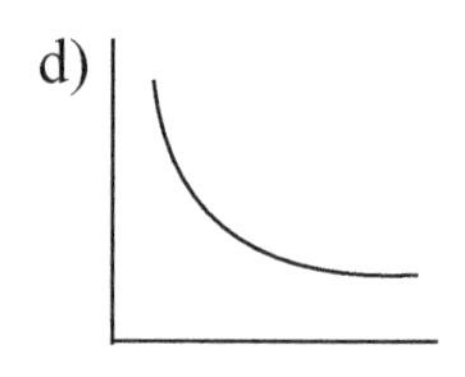

e) 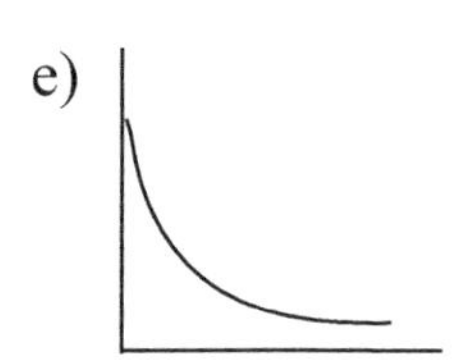

6. A plot of $t_{1/2}$ vs [A] for a reaction type aA → products which is zero order in A.

7. A plot of rate vs. [A] for a reaction type aA → products which is not zero-order and which occurs on the surface of a catalyst.

8. A plot of [A] vs. t for a reaction A → C which has the following mechanism

$$A \rightarrow B \quad \text{(slow)}$$
$$B \rightarrow C \quad \text{(fast)}$$

A Model for Chemical Kinetics and Catalysis

1. Provide a conceptual basis (at a molecular level) of zero-order, first-order, and second-order reactions. Why aren't all reactions second order if molecules must collide to react?

2. Adding a catalyst to a reaction system changes which of the following? Defend your answer.

   I. The mechanism of the reaction.
   II. The activation energy required to make the products.
   III. The temperature required to carry out the reaction at a reasonable rate.
   IV. The amount of products formed from a given amount of reactants.

The Equilibrium Condition and The Equilibrium Constant

1. Consider an equilibrium mixture consisting of $H_2O(g)$, $CO(g)$, $H_2(g)$, and $CO_2(g)$ reacting in a closed vessel according to the equation

$$H_2O(g) + CO(g) \rightleftharpoons H_2(g) + CO_2(g)$$

   a. You add more H_2O to the flask. How does the new equilibrium concentration of each chemical compare to its original equilibrium concentration after equilibrium is reestablished? Justify your answer.

   b. You add more H_2 to the flask. How does the concentration of each chemical compare to its original concentration after equilibrium is reestablished? Justify your answer.

2. Equilibrium is microscopically dynamic but macroscopically static. Explain what this means.

3. In Section 13.1 of your text, it is mentioned that equilibrium is reached in a "closed system". What is meant by the term closed system and why is it necessary for a system to reach equilibrium? Explain why equilibrium is not reached in an open system.

4. True or false? The amounts of the reactants and products are equal when equilibrium is reached. Explain your answer.

5. Differentiate between the terms equilibrium constant and equilibrium position. At a given temperature can a reaction have more than one value for an equilibrium constant? At a given temperature can a reaction have more than one equilibrium position?

6. Suppose we have the reaction as represented by the balanced equation

$$A(g) + B(g) \rightleftharpoons C(g)$$

   and at equilibrium, [A] = 2.0 *M*, [B] = 1.0 *M*, and [C] = 4.0 *M*. To a 1.0-L container of this system at equilibrium, you add 3.0 mol of B.

   Could the new equilibrium position have [A] = 1.0 *M*, [B] = 3.0 *M*, and [C] = 6.0 *M*? Note that in both cases $K = 2.0$.

7. When dealing with a gaseous system, we can use either pressure or concentration in the equilibrium expression. Use the equation PV=nRT to show how pressure and concentration are related.

8. As you saw in Section 13.1, generally adding a reactant shifts the equilibrium position to the right. But consider Figure 13.6 in your text. Adding $CaCO_3$ does not shift equilibrium. Why not?

9. Consider a reaction of all gases (all reactants and products are gases) for which $K = K_p$ at a given temperature.

   a. True or false? (choose the best response): Changing the volume at constant temperature will **not** shift the equilibrium position of the system.
      - True. There is no shift because a change in volume at constant temperature never shifts the equilibrium position.
      - True. There is no shift, but there would be a shift if $K \neq K_p$.
      - False. There will be a shift because a change in volume at constant temperature always shifts the equilibrium position.
      - False. There will be a shift, but there would not be a shift if $K \neq K_p$.

   b. Explain your answer to part a. As part of your explanation provide an example of a real reaction in which $K = K_p$, and an example of a real reaction in which $K \neq K_p$ and justify your choices. Full credit is reserved for explanations that use the example reactions and correctly incorporate partial pressures (and/or partial concentrations) of reactants and products.

Equilibrium Calculations

1. Being able to solve problems dealing with chemical equilibrium, it is a good idea to visualize what is happening at a molecular level. Problems 2 and 17 at the end of Chapter 13 in your text are good problems to consider in order to visualize equilibrium conditions. Do these problems.

2. These problems all give whole number answers so that you can concentrate less on the math and more on what is happening chemically.

   Consider the reaction represented by the equation

$$Fe^{3+}(aq) + SCN^{-}(aq) \rightleftharpoons FeSCN^{2+}(aq)$$

   a. In trial #1, you start with 6.00 *M* $Fe^{3+}(aq)$ and 10.0 *M* $SCN^{-}(aq)$ and at equilibrium the concentration of $FeSCN^{2+}(aq)$ is 4.00 *M*. What is the value of the equilibrium constant for this reaction?

   b. In trial #2, you start with 10.0 *M* $Fe^{3+}(aq)$ and 8.00 *M* $SCN^{-}(aq)$. What is the concentration of $FeSCN^{2+}(aq)$ at equilibrium?

   c. In trial #3, you start with 6.00 *M* $Fe^{3+}(aq)$ and 6.00 *M* $SCN^{-}(aq)$. What is the concentration of $FeSCN^{2+}(aq)$ at equilibrium?

   d. Find the equilibrium concentrations for all species for the following three trials:

	Fe^{3+}	SCN^{-}	$FeSCN^{2+}$
Trial #1:	9.00 *M*	5.00 *M*	1.00 *M*
Trial #2:	3.00 *M*	2.00 *M*	5.00 *M*
Trial #3:	2.00 *M*	9.00 *M*	6.00 *M*

3. What does the magnitude of the equilibrium constant, *K*, tell us?

4. In an ICE table, if the magnitude of *K* is large, does this mean *x* is small or large? Explain.

5. What is meant by the reaction quotient, *Q*? How does it compare to the equilibrium constant, *K*? That is, how is it similar, and how is it different?

6. Explain how we can use the value of the reaction quotient, *Q*, and the value of the equilibrium constant, *K*, to determine the direction a reaction will proceed.

7. What simplification in solving for equilibrium concentrations can we use when the value for *K* is very small? Explain.

8. Do the *Equilibrium Beads* activity. Directions for this activity are located at the end of this chapter.

LeChatelier's Principle

1. If the value of *K* for a reaction decreases as the temperature is increased, is the reaction exothermic or endothermic? Explain.

2. Consider a chemical system at equilibrium. The reaction is exothermic (releases energy as heat) as written and the temperature of the system is raised. Which way (if at all) does equilibrium shift, and what happens to the value of K? What if you add more product?

3. Consider an acidic chromate-dichromate system at equilibrium in which the color is an orange-yellow. The reaction is represented by the equation

$$\underset{\text{orange}}{Cr_2O_7^{2-}(aq)} + H_2O(l) \rightleftharpoons \underset{\text{yellow}}{CrO_4^{2-}(aq)} + H^+(aq)$$

   Which of the following best describes what happens if a strong base is added to the system at equilibrium? Explain your answer.

   a) Both the color of the solution and the value for *K* remain the same.
   b) The solution turns more yellow and the value for *K* increases.
   c) The solution turns more orange and the value for *K* decreases.
   d) The solution turns more orange and the value for *K* does not change.
   e) The solution turns more yellow and the value for *K* does not change.

4. For a system of gases at equilibrium, just knowing that there is a change in pressure does not tell us the direction of the shift in equilibrium or even if there is a shift. Find a chemical reaction for which an increase in pressure can shift equilibrium to the right, an increase in pressure can shift equilibrium to the left, and an increase in pressure does not affect equilibrium, depending on how the increase in pressure is attained.

Equilibrium Beads

Materials
100 type A (with stem) pop-it beads
100 type B (without stem) pop-it beads
Cardboard box reaction vessel
Blindfolds (2, optional)

Procedure
Assign roles to the different members of your group:

a. Forward reaction – the student finds two reactants and snaps them together to form a product molecule.
b. Reverse reaction – the student searches for product molecules and unsnaps or breaks the bonds to form reactants.
c. Agitator – the student agitates or shakes the cardboard reaction vessel to simulate the constant random kinetic motion of atoms and molecules.
d. Timer – the student times the reaction and starts and stops the activity.

1. Place 100 type A beads and 100 type B beads in the reaction vessel.
2. Blindfold the students representing the forward and reverse reactions (optional).
3. The Agitator begins shaking the reaction vessel. Caution: the box needs to be shaken ar a constant rate and care must be taken not to spill the beads.
4. The Timer signals the reactions to begin. The Forward Reaction Student forms only the product AB (not A_2). The Timer stops the reaction after one minute.
5. Count the number of products (AB) and the number of each reactant (A and B). Record these numbers.
6. Leave all products intact and repeat the process until equilibrium is established. Record the number of times required to reach equilibrium.

Summary Table

Time	Number of "A" particles	Number of "B" particles	Number of "AB" particles	Value for Q
Initial				
1 minute				
2 minutes				
3 minutes				
4 minutes				
5 minutes				
6 minutes				
7 minutes				
8 minutes				
9 minutes				
10 minutes				

Analysis and Conclusions

1. How can you tell when equilibrium is established?
2. Why must the agitator be careful to always shake the box at the same rate?
3. Compare your times required to reach equilibrium with those of other groups.
4. Compare your values of K with those determined by other groups.
5. Why shouldn't you expect the same value of K for each group? What causes different values?
6. How does this activity show that equilibrium is microscopically dynamic but macroscopically static?
7. List some differences between this activity and a chemical system achieving equilibrium.

The Nature of Acids and Bases

1. What is the equilibrium constant expression for an acid acting in water?

2. Consider the equation: $HA(aq) + H_2O \rightleftharpoons H_3O^+(aq) + A^-(aq)$

   a. If water is a better base than A^-, which way will equilibrium lie?
   b. If water is a better base than A^-, does this mean that HA is a strong or a weak acid?
   c. If water is a better base than A^-, is the value for K_a greater or less than 1?

3. It is a good idea to visualize solutions of weak and strong acids at a particulate level. Problem 6 and problem 44 at the end of Chapter 14 are problems specifically written to have you do this. Do these problems.

4. The autoionization of water, as represented by the below equation, is known to be endothermic. Which of the following correctly states what occurs as the temperature of water is raised? Defend your answer.

$$H_2O\ (l) + H_2O\ (l) \rightleftharpoons H_3O^+\ (aq) + OH^-\ (aq)$$

   a) The pH of the water does not change, and the water remains neutral.
   b) The pH of the water decreases, and the water becomes more acidic.
   c) The pH of the water decreases, and the water remains neutral.
   d) The pH of the water increases, and the water becomes more acidic.
   e) The pH of the water increases and the water remains neutral.

5. Choose the answer that best completes the following statement and defend your answer. When 100.0 mL of water is added to 100.0 mL of 1.00 *M* HCl,

   a) the pH decreases because the solution is diluted.
   b) the pH does not change because water is neutral.
   c) the pH is doubled because the volume is now doubled.
   d) the pH increases because the concentration of H^+ decreases.
   e) the solution is completely neutralized.

6. Choose the answer that best completes the following statement and defend your answer. Weak acid HA is weaker than weak acid HB. From this we know

   a) the pH of a solution of HA is lower than the pH of a solution of HB if these solutions are of equal concentration.
   b) the conjugate base A^- is weaker than the conjugate base B^-.
   c) the pH of at least one of the solutions must be greater than 7.00.
   d) the value of K_a for HA is lower than the value of K_a for HB.
   e) More information is required to answer this question.

7. You mix a solution of a strong acid with a pH of 4 and an equal volume of a strong acid solution with a pH of 6. Is the final pH (less than 4, between 4 and 5, 5, between 5 and 6, or greater than 6)? Explain.

8. You travel to a distant cold planct where the ammonia (NH_3) flows like water. In fact, the inhabitants of this planet use ammonia (an abundant liquid on their planet) like we use water. Ammonia is also similar to water in that it is amphoteric and undergoes autoionization. The value for equilibrium constant, K, for the autoionization of ammonia is 1.8×10^{-12} at the standard temperature of the planet. What is the pH of ammonia at this temperature?

9. You have two weak acids of equal concentration. Acid HA has a higher Ka value than acid HB. What does this mean? Which acid has the higher pH? Why?

Calculating the pH of Acid Solutions

1. What are the major species in a 2.0 *M* solution of HCl? What are the major species in a 2.0×10^{-3} *M* solution of HCl? Determine the pH of each solution at 25°C.

2. Determine the pH of a 1.5×10^{-11} *M* solution of HCl at 25°C.

3. The pH of a 1.0×10^{-7} *M* solution of HCl at 25°C is between 6.70 and 7.00. Why is this true?

4. Consider a solution of HCN.

   a. What are the major species?
   b. Of the two choices for the dominant reaction, which controls the pH? Explain.

   $HCN(aq) + H_2O \rightleftharpoons H_3O^+(aq) + CN^-(aq)$ $K_a = 6.2 \times 10^{-10}$

   $H_2O + H_2O \rightleftharpoons H_3O^+(aq) + OH^-(aq)$ $K_w = 1.0 \times 10^{-14}$

5. True or false? The pH of a solution of a strong acid is always lower than the pH of a solution of a weak acid. First, briefly explain your answer in complete sentences. Then, provide examples with real acids and use calculations to support your answer.

6. Answer the following without performing calculations.

   a. The value of K_a for a 4.00 *M* weak acid solution should be (higher/lower/the same) as the value of K_a of an 8.00 *M* solution of the same weak acid. Explain.
   b. The percent ionization of a 4.00 *M* weak acid solution should be (higher than/lower than /the same as) the percent ionization of an 8.00 *M* solution of the same weak acid. Explain.
   c. The pH of a 4.00 *M* weak acid solution should be (higher than/lower than/the same as) the pH of an 8.00 *M* solution of the same weak acid. Explain.

Bases

1. Acetic acid ($HC_2H_3O_2$) and HCN are both weak acids. Acetic acid is stronger than HCN. Arrange the following bases from weakest to strongest and explain your answer:

   H_2O Cl^- CN^- $C_2H_3O_2^-$

2. Consider the equation: $A^-(aq) + H_2O \rightleftharpoons HA(aq) + OH^-(aq)$, for which HA is a weak acid.
   a. Which way will equilibrium lie for the above equation?
   b. Is the value for K_b greater than or less than 1?
   c. Does this mean that A^- is a strong or a weak base?

3. The K_a value for acetic acid is 1.8×10^{-5}, and the K_a value for HCN is 6.2×10^{-10}. Calculate the K_b values for $C_2H_3O_2^-$ and CN^-.

4. Derive the relationship between K_a and K_b for a conjugate acid base pair.

5. If the acid-base reaction $HA + B^- \rightleftharpoons HB + A^-$ has a $K = 10^{-4}$, answer the following and explain:
   a. Which is the stronger acid, HA or HB?
   b. Which is the stronger base, A^- or B^-?

   How would your answers for the above change if $K = 10^4$? if $K = 1$?

6. Consider the equation $HCN(aq) + F^-(aq) \rightleftharpoons CN^-(aq) + HF(aq)$.
   a. Discuss whether the value of K for the following reaction is greater than, less than, or equal to 1, and explain how you would decide.
   b. Calculate the value for K for the equation.

7. True or false: A aqueous solution with a pH of 10.00 is basic, and therefore no H^+ is present in solution.

8. Indicate which of the graphs below best represents each plot described in a-d. Note: the graphs may be used once, more than once, or not at all.

a) 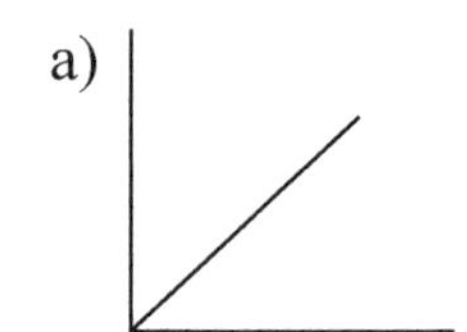

b) 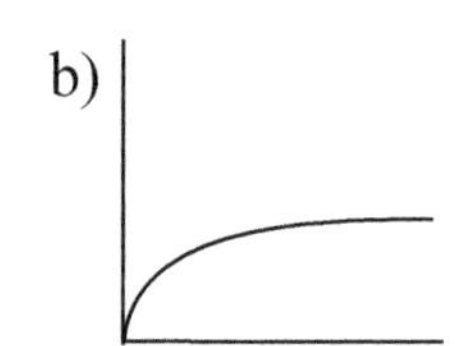

c) 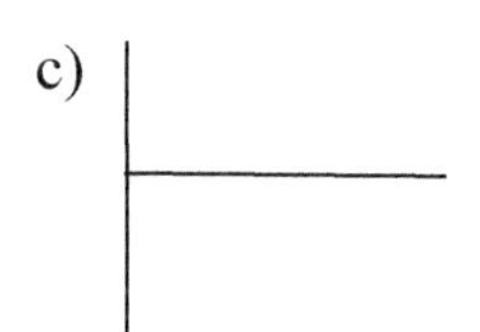

d) 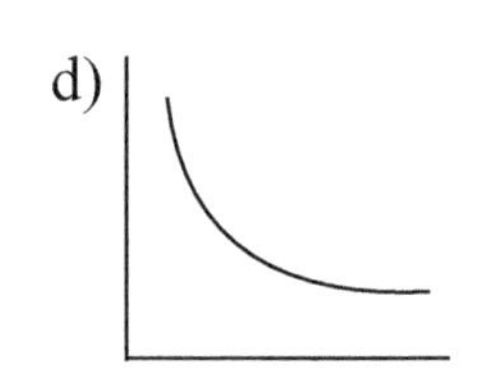

e) 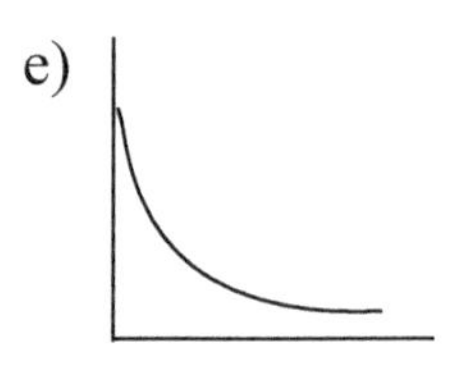

   a. Percent dissociation (y) vs. $[HA]_o$ (initial concentration) (x) for an aqueous weak acid at constant temperature.
   b. K_a (y) vs. K_b (x) for a series of aqueous conjugate acid-base pairs (constant T).
   c. $[H^+]$ (y) vs. $[HA]_o$ (initial concentration) (x) for an aqueous weak acid (constant T).
   d. $[H^+]$ (y) vs. $[OH^-]$ (x) for an aqueous solution at constant temperature.

Polyprotic Acids

1. Write the successive dissociations steps for the weak triprotic acid H_3A in water.

2. Consider Interactive Example 14.15 in your text. The conclusion is that for this case the second and third dissociation steps do not make an important contribution to $[H^+]$, and that the pH can be calculated by only considering the first dissociation step. Explain why this is true.

3. Consider Interactive Example 14.16 and Example 14.17 in your text. In one case you must consider the dissociation of HSO_4^- in determining the pH of the solution, and in the other case you do not. Explain why this is the case.

4. For which of the following 0.010 *M* diprotic acids would the second dissociation affect the pH significantly? Explain your answer.
   a) H_2A; $K_{a1} = 4.2 \times 10^{-2}$, $K_{a2} = 1.8 \times 10^{-7}$
   b) H_2B; $K_{a1} = 2.4 \times 10^{-4}$, $K_{a2} = 6.1 \times 10^{-8}$
   c) H_2C; $K_{a1} = 1.3 \times 10^{-4}$, $K_{a2} = 5.2 \times 10^{-9}$
   d) H_2D; $K_{a1} = 1.8 \times 10^{-3}$, $K_{a2} = 9.3 \times 10^{-4}$

5. For a general triprotic acid H_3A in water, the concentration of HA^{2-} at equilibrium will be about equal to the value of K_{a2} for acid. Why is this the case? Is the concentration of A^{3-} equal to the value of K_{a3} for the acid? Prove your answer.

Acid-Base Properties of Salts

1. True or false? The pH of any salt solution at 25°C is equal to 7.00. If true, explain why it is true and use a real example of a salt to support your answer. If false, provide a real example of a salt solution that has a basic pH and calculate its pH.

2. One of the difficulties with chemistry problems is trying to decide what information is important. Active Learning Question #14 at the end of Chapter 14 has you decide on information you need to solve a problem. Do this problem and provide a sample calculation.

3. To solve acid base problems you should consider what the solution "looks like" at a particulate level. We can often be given clues to the nature of a solution by measuring such factors as conductivity and pH. Problem 145 at the end of Chapter 14 requires you to put together many concepts and to visualize the solution. Do this problem.

4. Consider a 0.100 *M* aqueous solution of NaF. The K_a value for HF is 7.2×10^{-4}.
   a. What are the major species?
   b. Write all possibilities for the dominant reaction, and explain which controls the pH. Calculate the equilibrium constant for this reaction.
   c. Determine the pH of the solution of NaF.

5. Which if the following will raise the pH of a weak acid HA in aqueous solution? Defend your answer.
   I. Addition of water.
   II. Addition of NaA(*s*)
   III. Addition of NaCl(*s*).
   IV. Addition of HNO_3(*aq*)
   V. Addition of KOH(*aq*)

The Effect of Structure on Acid-Base Properties

1. In molecules of the type X-H, does the acid strength increase or decrease as the electronegativity of X increase? Explain and provide an example.

2. Explain how $Al(NO_3)_3$ dissolved in water can produce an acidic solution.

3. In molecules of the type X-O-H, as the electronegativity of X increases, the acid strength increases. In addition, if the electronegativity of X is quite low, the molecule acts like a base. Explain these observations and provide examples.

4. In your own words, explain the difference in acid-base properties of covalent oxides and ionic oxides. Provide examples.

The Lewis Acid-Base Model

1. Use the Lewis acid-base model to explain why covalent oxides in water form an acidic solution.

2. All Brønsted-Lowry acids and Arrhenius acids are also Lewis acids. Explain why this is true and provide examples.

3. Provide an example of a Lewis acid that is not an Arrhenius acid and explain your answer.

4. Provide an example of a Lewis base that is not a Brønsted-Lowry base and explain your answer.

5. Provide an example of a Brønsted-Lowry base that is not an Arrhenius base and explain your answer.

6. Provide an example of a Lewis base that is not an Arrhenius base and explain your answer.

Buffered Solutions

1. What is meant by the term *common ion effect*? Explain how it works.

2. Mixing together solutions of acetic acid and sodium hydroxide can make a buffered solution. Show how this is true and why it is dependent on the amounts of each.

3. Could a buffered solution be made by mixing aqueous solutions of HCl and NaOH? Explain.

4. You make 1.00 L of a buffered solution (pH = 5.00) mixing acetic acid ($K_a = 1.8 \times 10^{-5}$) and sodium acetate. You have 1.00 *M* solutions of each. What volume of each solution do you mix to make such a buffered solution? Explain why the relative volumes of the solutions make sense.

Titrations and pH Curves

1. Calculate the pH of an aqueous solution made by mixing 0.20 mol of $HC_2H_3O_2$ with 0.030 mol of NaOH in 1.0 L of aqueous solution.
   a. What are the major species?
   b. What are all possibilities for the dominant reaction?
   c. Which reaction controls the pH? How do you know?
   d. Solve for the pH.

2. You mix a solution of a strong acid with a pH of 4 and an equal volume of a strong base solution with a pH of 8. Is the final pH (less than 4, between 4 and 6, 6, between 6 and 8, or greater than 8)? Explain.

3. True or false? More base is required to bring 1.0 L of a 1.0 *M* solution of a strong acid to equivalence than to bring 1.0 L of a 1.0 *M* solution of a weak acid to equivalence. Defend your answer.

4. When titrating a strong acid with a strong base, why is the pH equal to 7.00 at the equivalence point?

5. What is true about the pH at the equivalence point when titrating a weak acid with a strong base? Defend your answer?

6. Why isn't the endpoint of a titration always equal to the equivalence point of a titration? Explain how to choose an indicator to keep this error to a minimum.

7. Use Figure 15.8 in your text to determine which indicators can you use when titrating a 100.0 mL solution of 0.100 M HF with 0.100 *M* NaOH. Provide calculations in your answer.

Solubility Equilibria

1. Sketch a graph of solubility (y) vs. K_{sp} (x) for a salt MX in pure water at constant temperature.

2. How does the solubility of silver chloride in water compare to that of silver chloride in an acidic solution (made by adding nitric acid to the solution)? Explain.

3. How does the solubility of silver phosphate in water compare to that of silver phosphate in an acidic solution (made by adding nitric acid to the solution)? Explain.

4. The salts AgX, AgY, and AgZ are all equally soluble in water (and none are very soluble at all). When each salt is added to separate beakers of 100 mL of 1.0 *M* HNO_3, you notice that AgY is much more soluble than AgZ in acid. The salt AgX is no more soluble in strong acid than it is in water. Rank the strength of the acids HX, HY, HZ and label each as strong or weak.

5. You have a saturated solution of aqueous NaCl (contains the maximum concentration of NaCl) and add 0.10 *M* HCl (*aq*) dropwise. Choose the statement below that best explains what will happen and defend your answer. The K_{sp} value for NaCl is equal to 38.

   a. The NaCl will precipitate out of solution and continue to do so as more 0.10 *M* HCl is added.
   b. At first NaCl will precipitate out of solution, but it will stop doing so after awhile. As more 0.10 *M* HCl is added, the NaCl solid will remain.
   c. At first nothing will seem to happen, but as more 0.10 *M* HCl is added, a precipitate will form.
   d. At first NaCl will precipitate out of solution, and then the NaCl will begin to re-dissolve. After awhile there will be no solid left.
   e. No NaCl will ever precipitate out of solution.

6. You mix two aqueous ionic solutions together and they do not react. To this you add (dropwise) a third aqueous ionic solution. At first a white solid forms. As you continue to add the third solution, a yellow solid begins to form. Which of the following can you say with confidence? Defend your answer.
   I. The solubility of the white solid is greater than the solubility of the yellow solid.
   II. The solubility of the yellow solid is greater than the solubility of the white solid.
   III. The K_{sp} value for the white solid is greater than the K_{sp} value for of the yellow solid.
   IV. The K_{sp} value for the yellow solid is greater than the K_{sp} value for of the white solid.

7. Problem 6 above requires you to compares solubilities and K_{sp} values of possible products. Active Learning Question 5 at the end of Chapter 16 in your text also requires this. Do this problem.

8. Consider Problem 79 at the end of Chapter 16 in your text. First, answer the question as it is written. Then, consider that the NH_3(*aq*) added is concentrated. Initially you will see the same result as test tube 2. Upon further addition of concentrated NH_3(*aq*), a white solid will be seen, as in test tube 3 (although the nature of the solid is different). Explain this.

Spontaneous Processes and Entropy

1. Consider 2.4 moles of a gas contained in a 4.0-L bulb at a constant temperature of 32°C. This bulb is connected to an evacuated 20.0-L sealed bulb via a valve. Assume that the temperature remains constant.
   a. What should happen to the gas when you open the valve?
   b. Calculate ΔH, ΔE, q, and w for the process you described above.
   c. Given your answer to part b, what is the driving force for the process?

2. According to your text, what is the driving force for a process to be spontaneous? Explain your answer.

3. Predict the sign of ΔS for each of the following, and explain:
   a. the evaporation of alcohol
   b. the freezing of water
   c. compressing an ideal gas at constant temperature
   d. heating an ideal gas at constant pressure
   e. dissolving NaCl in water

4. Is ΔS_{univ} a state function? Explain.

5. What does the sign ΔS_{surr} of tell us? Explain your answer.

6. What does the magnitude ΔS_{surr} of tell us? Explain your answer.

7. Describe the following as spontaneous/non-spontaneous/cannot tell, and explain. Also, provide a real example for each. A reaction that is
   a. exothermic and becomes more positionally random
   b. exothermic and becomes less positionally random
   c. endothermic and becomes more positionally random
   d. endothermic and becomes less positionally random

   Explain how temperature affects your answers.

8. Use the ideas of energy randomness and positional randomness in your discussion of the following questions.
   a. Under what conditions is the freezing of water spontaneous? Why?
   b. Under what conditions is the melting of ice spontaneous? Why?
   c. Under what conditions is the freezing of water as likely as the melting of ice? Why?

9. From the following, choose the one that best completes the following statement, and defend your answer. The dissociation of hydrogen gas as described by the equation

$$H_2(g) \rightleftharpoons 2H(g)$$

   a) is spontaneous at any temperature.
   b) is spontaneous at low temperatures.
   c) is spontaneous at high temperatures.
   d) is spontaneous at only one specific temperature.
   e) never takes place.

10. At least some of what is in the following quoted statements is false.

    "By *spontaneous* we mean that the reaction or process will always proceed to the right (as written) even if very slowly. Increasing the temperature may speed up the reaction, but it does not affect the spontaneity."

    What is right in the statements above? What is wrong? Provide a real world example of temperature dependent spontaneity for a process or reaction. Explain the relevant driving forces, and how temperature changes the outcome of the process/reaction.

11. Not all exothermic reactions or process are spontaneous. However, the fact that a reaction or process is exothermic is a driving force for the reaction/process to occur.

    a. Why is exothermicity a driving force for a reaction? Discuss how exothermicity can lead to an increase in ΔS_{univ}.
    b. If exothermicity is a driving force, why aren't all exothermic reactions or processes spontaneous? That is, is being exothermic causes a reaction/process to proceed, what "stops" it? Provide a real example and discuss it.
    c. How does temperature affect the significance of exothermicity as a driving force? That is, if an exothermic reaction/process is run at a higher temperature, does the fact that it is exothermic matter more, less, or the same in determining spontaneity? Explain your answer and provide a real example.

Free Energy

1. Rubidium has a heat of vaporization of 69.0 kJ/mol at its boiling point (686°C). Consider the process

$$Rb(l) \rightarrow Rb(g) \quad (1 \text{ atm}, 686°C)$$

    a. Calculate ΔS (in J/K mol).
    b. Calculate ΔG (in kJ/mol).

2. The value of $\Delta H_{vaporization}$ for substance X is 45.7 kJ/mol, and its normal boiling point is 72.5°C. Calculate ΔS, ΔS_{surr}, and ΔG for the vaporization of 1 mole of this substance at 72.5°C, and 1 atm.

3. A stable diatomic molecule spontaneously forms from its atoms at constant temperature and pressure. Predict the signs of: $\Delta H°$, $\Delta S°$, and $\Delta G°$. Explain your answers.

4. Why do chemical reactions reach equilibrium? That is, if products are more stable than reactants, why doesn't the reaction proceed to completion? Conversely, if reactants are more stable than products, why does the reaction proceed at all?

5. Consider the freezing of liquid water at –10°C and 1 atm. For this process determine the signs for ΔH, ΔS, and ΔG and defend your answer.

6. A mixture of hydrogen and chlorine remains unreacted until it is exposed to ultraviolet light from a burning magnesium strip. Then the following reaction occurs very rapidly. Which best explains this? Defend your answer.

$$H_2(g) + Cl_2(g) \rightarrow 2HCl(g) \qquad \Delta G = -45.54 \text{ kJ}, \quad \Delta H = -44.12 \text{ kJ}, \quad \Delta S = -4.76 \text{ J/K}$$

   a) The reactants are thermodynamically more stable than the products.
   b) The reaction has a small equilibrium constant.
   c) The ultraviolet light raises the temperature of the system and makes the reaction more favorable.
   d) The negative value for ΔS slows down the reaction.
   e) The reaction is spontaneous, but the reactants are kinetically stable.

7. Of the following choices, choose the one which best completes the following statement and defend your answer. For the vaporization of a liquid at a given pressure: (Note: Answer c and answer d each imply that ΔG is zero at some temperature.)

   a) ΔG is positive at all temperatures.
   b) ΔG is negative at all temperatures.
   c) ΔG is positive at low temperatures, but negative at high temperatures.
   d) ΔG is negative at low temperatures, but positive at high temperatures.
   e) The answer depends on the nature of the liquid.

8. A liquid is vaporized at its boiling point. Determine the sign for each of the following thermodynamic functions and defend your answer.

   q, w, ΔS, ΔS_{surr}, ΔS_{univ}, ΔH, ΔG

9. When 50.0 g of ice at 0°C is dropped into 100.0 g of water at 75°C in a perfectly insulated vessel all of the ice will melt and the final temperature of the 150.0 g of water will be above 0°C. Determine ΔS and ΔS_{univ} for this process and discuss how it shows this process to be spontaneous. You will need to determine ΔS and the final temperature of the water.

   ΔH_{fusion} for ice is 6.02 kJ/mol at 0°C and C_p for water is 75.3 J/Kmol

   Assume C_p is constant over the temperature range in question.

Galvanic Cells and Standard Reduction Potentials

1. Why is a systematic method necessary for balancing oxidation-reduction reactions? Why can't these equations be balanced by inspection?

2. We use half-reactions because they make balancing oxidation-reduction equations easier. However, half reactions do not occur in nature. Why not?

3. In balancing oxidation-reduction equations, why is it permissible to add water to either side of the equation?

4. What does it mean for a substance to be *oxidized*? The term oxidation originally came from substances reacting with oxygen gas. Explain why a substance which reacts with oxygen gas will always be oxidized.

5. Active Learning Question 3 at the end of Chapter 18 asks you to consider how a galvanic cell works. Do this problem.

6. Sketch a cell made with the following solutions and electrodes, and calculate the potential of each cell. Include the direction of electron flow in each cell, and label the anode and the cathode.
   a. Ag electrode and 1.0 *M* $Ag^{+}(aq)$ with Cu electrode and 1.0 *M* $Cu^{2+}(aq)$.
   b. Zn electrode and 1.0 *M* $Zn^{2+}(aq)$ with Cu electrode and 1.0 *M* $Cu^{2+}(aq)$.

   What should happen to the potential if you increase the $[Cu^{2+}]$ in the cell for part b? Explain.

7. A galvanic cell is constructed from two different metals and ions of these metals (assume the solutions each have a concentration of 1.0 *M*). To determine the standard cell potential, you must reverse one of the reactions as written. Why must we reverse one of these reactions and how do you know which one of these to reverse? Also, provide two examples from Table 18.1 in your text (in one example, the signs of the standard reduction potentials should be the same, in the other, the signs should be different). Determine the standard potential of the cells in both cases (show all work).

8. According to your text, "The value of $\mathcal{E}°$ is not changed when a half-reaction is multiplied by an integer". Why do we not multiply the potentials of half reactions when we multiply the half reactions? Provide conceptual and algebraic support.

9. Problem 160 at the end of Chapter 18 in your text asks you to create a table to standard reduction potentials given data. This is a good way to show understanding of what the reduction potentials tell us. Do this problem.

Thermodynamics and Dependence of Cell Potential on Concentration

1. For a reaction in a galvanic (voltaic) cell $\Delta S°$ is negative. Which of the following statements is true? Defend your answer.
   a) $\mathcal{E}$ will increase with an increase in temperature.
   b) $\mathcal{E}$ will decrease with an increase in temperature.
   c) $\mathcal{E}$ will not change when the temperature increases.
   d) $\Delta G° > 0$ for all temperatures.
   e) None of the above statements is true.

2. Given the following two standard reduction potentials,

   $M^{3+} + 3e^- \rightarrow M$ -0.10 V
   $M^{2+} + 2e^- \rightarrow M$ -0.50 V

   Solve for the standard reduction potential of the following half-reaction

   $M^{3+} + e^- \rightarrow M^{2+}$

3. You make a galvanic cell with a piece of nickel, 1.0 *M* $Ni^{2+}(aq)$, a piece of silver, and 1.0 *M* $Ag^+(aq)$.
   a. Sketch a diagram of this cell, labeling the anode and cathode, and showing the direction of the electron flow. Then calculate the cell potential.
   b. Calculate the concentrations of $Ag^+(aq)$ and $Ni^{2+}(aq)$ once the cell is "dead."

Corrosion and Electrolysis

1. If iron is left in the air, it will rust (corrode). Aluminum, like iron, is a reactive metal. However, aluminum generally does not corrode severely. Explain why not.

2. An unknown metal (M) is electrolyzed. It took 74.1 sec for a current of 2.00 amp to plate 0.107 g of the metal from a solution containing $M(NO_3)_3$. What is the metal?

3. Consider a solution containing 0.10 *M* of each of the following: Pb^{2+}, Cu^{2+}, Sn^{2+}, Ni^{2+}, and Zn^{2+}. In what order are the metals expected to plate out as the voltage is turned up from zero? Do the metals form on the cathode or on the anode? Prove it.

4. Problem 138 at the end of Chapter 18 in your text is a good problem to consider how we can use electrolysis to separate metals from solution. Do this problem.

Radioactive Decay and Nuclear Transformation

1. It states in Section 19.1 of your text that there are about 2000 nuclides known. How can this be possible if there are only about 100 elements?

2. Write the charge, mass number, and symbol for a beta particle and an alpha particle.

3. By how many units does the mass number of a nucleus change when the nucleus produces an alpha particle? By how many units does the mass number of a nucleus change when the nucleus produces a beta particle? Is each change an increase or decrease in mass number?

4. Is nuclear decay a zero-order, first-order, or second-order process? Why?

5. Determine the number of half-lives that must pass for only 1% of a particular radioisotope to remain.

6. Problem 71 at the end of Chapter 19 is a good problem to test your understanding of the concept of a half-life. Do this problem.

7. Do the *Half-Life of Pennies* activity. You can find the procedure at the end of this chapter.

8. Hundreds of years ago, alchemists tried to turn lead into gold. Is this transformation possible? If not, why not? If yes, how would you do it?

Radiation and Energy Changes

1. Do radiotracers generally have long or short half-lives? Why?

2. Why is the nucleus of an atom a great source of energy?

3. Discuss what is meant by the term critical mass and explain why the ability to achieve a critical mass is essential to sustaining a nuclear reaction.

4. What makes fusion preferable over fission? What makes fusion more complicated?

5. Why is it difficult to determine the effects of radiation on humans?

The Half-Life of Pennies

Materials

100 pennies
Graph paper

Procedure

1. Flip 100 pennies and separate them according to which landed heads and which landed tails. Record the number of heads.

2. Flip only the pennies that landed heads, and then separate the pennies according to which landed heads and which landed tails. Record the number of heads. Repeat this until you are out of pennies. Record the number of times until you are out of pennies.

Analysis and Conclusions

1. In this experiment, what do the pennies that land "heads" represent?
2. In this experiment, what do the pennies that land "tails" represent?
3. In this experiment, what does each flip represent?
4. Make a graph of number of pennies flipped vs. trial number from your data.
5. Gather together all of the class data and make a second graph of the total number of pennies flipped vs. trial number.
6. Why is there a difference between the graph of your data and graph of the class data?
7. Draw a graph that shows the decay of a 100.0-g sample of a radioactive nuclide with a half-life of 10 years. This should be a graph of mass versus time for the first four half-lives.
8. Compare the two graphs using your data and the class data to the graph of the 100.0 g sample. Does your graph or the graph of the class data look more like the graph of the 100.0-g sample? Why?
9. Approximately how many half-lives would it take for one mole of a radioactive nuclide to completely disappear?

The Transition Metals and Coordination Compounds

1. What is the expected electron configuration of Sc^+? Explain.

2. A coordinate covalent compound is dissolved in water and an excess of $AgNO_3$ is added (rccall that AgCl is relatively insoluble in water). If 1.0 mol of each of the following compounds is dissolved in water, the addition of excess $AgNO_3$ to which one would give rise to the greatest number of moles of solid AgCl? Defend your answer.
   a) $[Co(H_2O)_6]Cl_2$
   b) $K_4[CoCl_6]$
   c) $[Co(H_2O)_5Cl_2]Cl$
   d) $K_2[CoCl_4]$
   e) $[Co(H_2O)_3Cl_3]$

3. What is the difference between the *primary valence* and *secondary valence* for a coordination compound? Provide an example coordination compound and point out each of the valences.

4. What is a ligand? How does it bind to a metal ion in a coordination compound?

5. How many isomers of $[Co(en)_2Cl_2]Cl$ are there? Draw them, and list the type of isomer for each.

6. Which of the following compounds requires the *cis*/*trans* designation when naming? Defend your answer.
   a) pentaamminebromocobalt(II) chloride
   b) diamminedichlorocobalt(II) (tetrahedral)
   c) diamminedichloroplatinum(II) (square planar)
   d) $[Co(en_3)]Cl_2$

7. How many octahedral coordinate covalent isomers with the partial name "tetraaqua…" can be written with the formula $CoF_2Cl{\bullet}4H_2O$?

The Crystal Field Model

1. Consider the crystal field model.
   a. Which is lower in energy, lobes pointing to ligands or between ligands? Why?
   b. Consider the electrons in the *d*-orbitals in the crystal field model. Are they from the metal or the ligands?
   c. Why would electrons pair up in *d*-orbitals instead of being in separate orbitals?
   d. Why would you predict that the splitting in tetrahedral complexes is smaller than that in octahedral complexes?

2. Sketch a crystal field diagram for an octahedral complex and explain/justify the relative positions of the *d*-orbitals (make sure to label these).

3. Use the Crystal Field Theory to explain the colors of transition metal ions in solution. Why are there colors, and why/how are there different colors?

4. How many of the following are expected to form colorless octahedral compounds? Explain

Zn^{2+}, Fe^{2+}, Mn^{2+}, Cu^{+}, Cr^{3+}, Ti^{4+}, Ag^{+}, Fe^{3+}, Cu^{2+}, Ni^{2+}

5. Which of $[Mn(CN)_6]^{3-}$ and $[Mn(CN)_6]^{4-}$ is more likely to be high spin? Why?

6. The complex ion $[ML_6]^{2+}$ is found to be diamagnetic, and the ligand L is known to be neutral. Provide a metal that could be M and defend your answer.

7. Consider the complex ions $[Co(A)_6]^{2+}$, $[Co(B)_6]^{2+}$, $[Co(A)_6]^{3+}$, and $[Co(C)_6]^{3+}$. The ligands A, B, and C are all neutral. The numbers of unpaired electrons for each of the complex ions (respectively) are found to be: 3, 1, 0, and 4. Rank the ligands from strong field to weak field.

8. When concentrated hydrochloric acid is added to the red solution $[Co(H_2O)_6]^{2+}$, the blue $[CoCl_4]^{2-}$ (tetrahedral) forms. Explain this color change.

9.

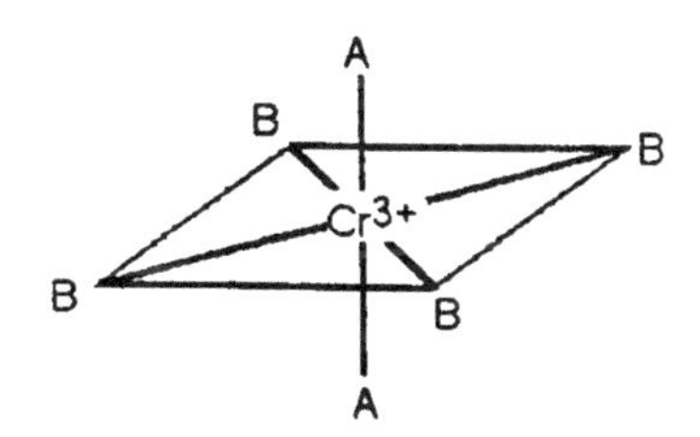

Consider the pseudo-octahedral complex of Cr^{3+} shown above, where A and B represent neutral ligands. Draw an appropriate crystal field diagram for this complex (include the electrons). Explain (complete sentences) and fully support your answer using the ideas of the crystal field theory. If you cannot choose between what you believe to be equally good diagrams, explain why.

Note: The $[CrA_6]^{2+}$ complex ion has 2 unpaired electrons; the $[CrB_6]^{2+}$ complex ion has 4 unpaired electrons.

10. A metal ion in an octahedral coordination compound has 4 more unpaired electrons in the high-spin case than in the low-spin case. Find a metal ion that fits this description and defend your answer.

11. You discover a new ligand and wish to place it on the spectrochemical series. Which of the following ions would be a bad choice to determine if the ligand were a weak field or strong field ligand? Defend your answer.
 a) Cr^{2+}
 b) Co^{2+}
 c) Ni^{2+}
 d) Fe^{2+}

Hydrocarbons

1. What is meant by the term "unsaturated hydrocarbon"? What structural feature characterizes unsaturated hydrocarbons?

2. The following are named incorrectly, but a correct structure can be made from the name given. Draw the following incorrectly named compounds and name them correctly:
   a. 2-ethyl-3-methyl-5-isopropyl hexane
   b. 3-methyl-4-isopropylpentane
   c. 2-ethyl-3-butyne

3. Which of the following names is a correct one? Defend your answer.
   a) 3,4-dichloropentane
   b) 1,1-dimethyl-2,2-diethylbutane
   c) *cis*-1,3-dimethylpropane
   d) 2-bromo-1-chloro-4,4-diethyloctane
   e) At least two of the above are correct.

4. How many isomers are there for C_5H_{12}? Draw their structures and name them.

5. When propane undergoes dehydrogenation, what is the product? Name it and draw a structural formula.

6. Which of the following molecules exist? Draw structural formulas to make your decision.
   a. methene
   b. 5,5,5-trichloro-5-bromo-2-pentene
   c. cycloethane
   d. 2,2-dichloropropene

7. Which of the following is an incorrect name? Defend your answer.
   a) *trans*-1,2-dichloroethene
   b) propylene
   c) ethylene
   d) *cis*-1,2-dichloroethane

8. Explain what is incorrect about the names given below, and provide a correct name for both.
   a. 3-pentyne
   b. 1-propyne

Functional Groups

1. How many different possible "tetramethylbenzenes" exist? Provide structural formulas and names for each.

2. How is the benzene ring named if it is considered a substituent in another molecule? Give the structures and names of two examples.

3. If a benzene ring contains several substituents, how are the relative locations of the substituents numbered in the systematic name for the molecule? Provide an example.

4. Match the molecules with the boiling points below and explain your answer.

Molecules	boiling points
ethanol	-42.1°C
propane	-23°C
dimethyl ether	78.5°C

5. One of your friends names a compound 1-propanone. Name this compound correctly.

6. For which of the following compounds are *cis* and *trans* isomers possible? Defend your answer.
   a) 2,3-dimethyl-2-butene
   b) 3-methyl-2-pentene
   c) 4,4-dimethylcyclohexanol
   d) ortho-chlorotoluene

7. For the general formula $C_6H_{14}O$, draw the structures of three isomeric alcohols that illustrate primary, secondary, and tertiary structures.

8. Name and draw the structure of the principal organic product when each of the following is oxidized.
   a. 1-propanol
   b. 2-butanol
   c. 2-methy-2-propanol

Polymers

1. Differentiate between an addition polymer and a condensation polymer and provide an example of each. Do the same for copolymer and homopolymer.

2. Draw the structure of the monomer and the basic repeating unit for each of the following polymeric substances. Also, provide one use for each of the polymers.
   a. polyvinyl chloride
   b. Teflon
   c. polystyrene

3. Differentiate among primary, secondary, and tertiary structure in proteins.

4. Draw the structures of the simple dipeptides *gly-ala* and *ala-gly*.

5. How are proteins able to provide a buffering action?

6. What is meant by inhibition of an enzyme? What happens when an enzyme is irreversibly inhibited?

7. Differentiate between monosaccharide and disaccharide. Provide an example of each.

8. Sketch a representation of sucrose and clearly label the portion that originates from glucose, the portion that originates from fructose, and the glycoside linkage between the rings.

9. What is a polysaccharide? What monomer unit makes up starch and cellulose?

10. Describe the structure of a typical nucleotide.

11. Sketch the structures of the sugars ribose and deoxyribose. Which molecule, RNA or DNA, contains each sugar?

12. Sketch the general structure of a triglyceride. What are the components that go into making a typical triglyceride?

13. Describe the mechanism by which fatty acid salts are able to exert a cleaning action.